Praise for *Feral*

"Drawing on a life of rich observation and experience, George Monbiot regales us with stories of life's astonishing capacity for renewal and offers an uplifting and inspiring goal beyond the cessation of our destructive rampage—the restoration of the wild in nature and our own lives." —David Suzuki

"It could not be more rigorously researched, more elegantly delivered, or more timely. We need such big thinking for our own sakes and those of our children. Bring on the wolves and whales, I say, and, in the words of Maurice Sendak, let the wild rumpus start." —Philip Hoare, *Sunday Telegraph*

"The world knows George Monbiot mostly from his powerful and perceptive journalism. But this is a whole different order of writing and thinking, a primal account of an unstifled world." —Bill McKibben, author of *The End of Nature* and *Eaarth*

"Monbiot is a proper reporting journalist, he can write, and he stands for something—which puts him, these days, well ahead of most of our tribe. Plus, this peculiar and involving book—three-quarters exhilarating environmental manifesto, one quarter midlife crisis—has an enormous amount to recommend it. . . . Extraordinarily good and crunchy material. . . . There's a lot here to digest and think about, much to be excited by." —*The Spectator*

"A highly analytical and richly researched book." —*Maclean's*

"In this remarkable book, the journalist and environmentalist George Monbiot explores projects where this 'incendiary idea' has been put into practice. The results are extraordinary. . . . Most impressive about *Feral* is its focus on finding constructive solutions to ecological problems." —*Sunday Times*

"Monbiot's book is wadded full with stories and facts aplenty, but the quality that most endures are his descriptions of the bigger world. . . . The tangible, almost perfume-heavy descriptions of the landscape and the creatures that inhabit them are wondrous and dream-like. Cinematic." —*The Tyee*

"*Feral* has really opened my mind to the history and possibilities of our landscape. It reflects a very real need in us all right now to be released from our claustrophobic monoculture and sense of powerlessness. To break the straight lines into endless branches. To free our land from its absent administrators. To rewild both the landscape and ourselves. It is the most positive and daring environmental book I have read. In order to change our world you have to be able to see a better one. I think George has done that." —Thom Yorke of Radiohead

"A fun bit of investigative journalism. . . . [Monbiot] is a gifted nature writer." —*Toronto Star*

"Monbiot has the visionary polemicist's gift of pursuing an argument by gentle stages to a dazzlingly aspirational conclusion. His accounts of the ecological horrors perpetrated by sheep and the perverse defence of their depredations by assorted conservation bodies are not just persuasive but powerfully affecting. He is brilliant, too, at presenting statistics in readable form, and on the adroitly irrefutable deployment of ancient historical evidence. . . . Something about the charm and persistence of Monbiot's argument has the hypnotic effect of a stoat beguiling a hapless rabbit. Soon you find yourself dazedly agreeing that it's all a tremendous idea." —*New Statesman*

"To read this seminal, subversive, sometimes intoxicating book could mean never to look at our landscape in quite the same way again. . . . *Feral* belongs on the shelf with Roger Deakin, Richard Mabey, Robert Macfarlane, Kathleen Jamie and other fine writers who have engaged in the human reunion with nature." —*The Irish Times*

"Monbiot's latest book stands in a long tradition of back-to-nature narratives, the most famous of which is Thoreau's *Walden*. It is also, at one level at least, a mid-life crisis memoir. However, *Feral* is both more original and more important than such a description would suggest. . . . Wolves, he tells us, are 'necessary monsters of the mind'; perhaps the same could be said of Monbiot himself."

—*The Independent*

"There's nothing ignoble about Monbiot's vision of reinstating ecosystems in which man's power to dominate is consciously withheld. It is a vision fed by his growing disenchantment with the landscape that surrounds him. . . . Rewilding along the lines Monbiot advocates becomes an attractive proposal, a hopeful metaphor for something over nothing."

—*The Guardian*

"Part personal journal, part rigorous (and riveting) natural history, but above all unbridled vision for a less cowed, more self-willed planet, this is a book that will change the way you think about the natural world, and your place in it. Big, bold and beautifully written, his vision of a rewilded world is, well, truly captivating."

—Hugh Fearnley-Whittingstall, celebrity chef
and author of *The River Cottage Cookbook*

"A Book of Revelations for our times. It warns us in no uncertain terms that if we don't change our ways in the hell of a hurry, we'll have done two other things: 1) Committed the ultimate crime of biocide; and 2) Hanged ourselves in the process thereof. Read *Feral* and act . . . or else."

—Farley Mowat, author of *Sea of Slaughter*
and *Never Cry Wolf*

"George Monbiot is always original—both in the intelligence of his opinions and the depth and rigour of his research. In this unusual book he presents a persuasive argument for a new future for the planet, one in which we consciously progress from just conserving nature to actively rebuilding it."

—Brian Eno

GEORGE MONBIOT

Feral

Rewilding the Land, the Sea and Human Life

THE UNIVERSITY OF CHICAGO PRESS • CHICAGO

GEORGE MONBIOT studied zoology at Oxford, but his real education began when he travelled to Brazil in his twenties and joined the resistance movement defending the land of indigenous peasants. Since then he has spent his career as a journalist and environmentalist, working with others to defend the natural world he loves. His celebrated *Guardian* columns are syndicated all over the world. Monbiot is the author of the books *Captive State*, *The Age of Consent*, *Bring on the Apocalypse*, and *Heat*, as well as the investigative travel books *Poisoned Arrows*, *Amazon Watershed*, and *No Man's Land*. Among the many prizes he has won is the UN Global 500 award for outstanding environmental achievement, presented to him by Nelson Mandela.

The University of Chicago Press, Chicago 60637

The University of Chicago Press, Ltd., London

© 2014 by George Monbiot

All rights reserved. Published 2014.

Printed in the United States of America

23 22 21 20 19 18 17 16 15 14   1 2 3 4 5

ISBN-13: 978-0-226-20555-7 (cloth)

ISBN-13: 978-0-226-20569-4 (e-book)

DOI: 10.7208/chicago/9780226205694.001.0001

LIBRARY OF CONGRESS CATALOGING-IN-PUBLICATION DATA
Monbiot, George, 1963– author.
  Feral : rewilding the land, the sea, and human life / George Monbiot.
  pages cm
  Includes index.
  ISBN 978-0-226-20555-7 (cloth : alkaline paper) — ISBN 978-0-226-20569-4 (e-book) 1. Wildlife reintroduction. 2. Restoration ecology. I. Title.
  QL83.4.M66 2014
  639.97'9—dc23

                                        2014013971

♾ This paper meets the requirements of
ANSI/NISO Z39.48-1992 (Permanence of Paper).

To Rebecca, Hanna and Martha
With all my love

And in memory of Morgan Parry,
an honest man

Feral: *'in a wild state, especially after escape from captivity or domestication'*

Contents

Preface

Arrange these threats in ascending order of deadliness: wolves, vending machines, cows, domestic dogs and toothpicks. I will save you the trouble: they have been ordered already.

The number of deaths known to have been caused by wolves in North America during the twenty-first century is one[1,2]: if averaged out, that would be 0.08 per year. The average number of people killed in the US by vending machines is 2.2 (people sometimes rock them to try to extract their drinks, with predictable results).[3] Cows kill some twenty people in the US,[4] dogs thirty-one.[5] Over the past century, swallowing toothpicks caused the deaths of around 170 Americans a year.[6] Though there are sixty thousand wolves in North America, the risk of being killed by one is almost nonexistent.

If you find that hard to believe, you are not alone, and not to blame. For centuries we have terrified ourselves with tales of the lethal threat wolves present to humankind, and the unending war being fought with equal vigor on both sides. In reality, wolves are

1. Candice Berner, in Alaska on 8 March 2010. http://www.adfg.alaska.gov/static/home/news/pdfs/wolfattackfatality.pdf.
2. The cause of a second death, that of Kenton Joel Carnegie in Saskatchewan, Canada, in 2005, is disputed. The evidence appears to suggest that it is more likely that he was killed by a bear.
3. http://urbanlegends.about.com/b/2005/06/29/are-vending-machines-deadlier-than-sharks-repost.htm.
4. http:/tierneylab.blogs.nytimes.com/2009/07/31/dangerous-cows/?_php=true&_type=blogs&_r=0.
5. http://historylist.wordpress.com/2008/05/29/human-deaths-in-the-us-caused-by-animals/.
6. http://www.videojug.com/interview/unlikely-ways-to-die#how-many-people-have-died-from-toothpicks.

exceedingly afraid of people and in almost all circumstances avoid us. If we take the time to win their trust—as the biologists who have been adopted by wild wolf packs can testify—they can become affectionate companions. But the fairytales are more powerful than the facts.

Could it be that we are so afraid of wolves not because they represent an alien threat, but because we recognize in them some of our own traits? They have a similar social intelligence: the ability to interpret and respond to someone else's behavior and mood. They look at you as if they can read your mind. To some extent they can, which is why we domesticated them. This, perhaps, is why they unnerve us, and why so many stories have been written and filmed in which wolves become humans or disguise themselves as such, or humans become wolves.

But perhaps there is something else at work too, a subliminal yearning for the kind of danger that no longer infects our lives. Discovery Channel's very popular series *Yukon Men* is as accurate a description of the world as the tales of Jacob and Wilhelm Grimm. It claims that 'there have been twenty fatal wolf attacks in the last ten years.'[7] This would be wrong under any circumstances, but the strong implication is that all these attacks have taken place in and around the town of Tanana. This town, the series tells us, 'is under siege by hungry predators . . . there's always somebody that's not going to make it home.'

Amid scenes of revolting cruelty inflicted by hunters clumsily killing (or trying to kill) the animals they have caught in their traps, the series insists that the men have no choice: otherwise these animals would stalk and gut them. Even wolverines, it says, 'are capable of tearing human beings apart.' When the biologist Adam Welz investigated this claim, he was unable to find a documented case of a wolverine attacking anyone, anywhere on earth.[8] Had the series maintained that the town was being stalked by killer vending machines, the claim would have been no less plausible.

Programs of this kind now throng the television schedules. Discovery

7. Adam Welz, 17 May 2013, 'Bloodthirsty "factual" TV shows demonize wildlife', http://www.theguardian.com/environment/nature-up/2013/may/17/ bloodthirtsty-wildlife-documentaries-reality-ethics.
8. Adam Welz, as above.

has also broadcast a chilling documentary which claims that *Carcharo-don megalodon*, a giant shark which has been extinct for over a million years, is still alive and roaming the oceans. In support of this thesis, it shows the horribly mutilated carcass of a whale, washed up on a beach.[9] A contributor tells us "you can clearly see a bite radius in the whale. . . . The whale looks to be almost bit in half, it's absolutely insane. Local marine biologists analysed the whale and determined—as crazy as it sounds—that the tail was bitten off in one bite." That the picture looks like a clunky computer-generated image appears to be no deterrent to the thesis; or to the fabulous viewing figures.[10]

The success of these shows reinforces the notion that we wish to believe we are surrounded by ancient terrors. Like the thousands of annual sightings of imaginary big cats, the ratings suggest we are missing something—something rich and grand and thrilling which resonates with our evolutionary history. Our imagination responds vividly to threats of the kind that we evolved to avoid. In the absence of sabretooths, lions and rampaging elephants, wolves will have to do.

So when I see the myths propounded by *Yukon Men* or by the organizers of the Salmon Predator Derby, which encouraged people to travel to Salmon, Idaho, and compete for a prize of $1,000 for killing the largest wolf,[11] I wonder whether some people hate wolves for the same reason that others love them. Because they have come to embody the fear and thrill that is often missing from our lives, people will fight to re-establish them as fiercely as others will fight to exterminate them.

Nowhere are these conflicts played out with greater intensity than in North America. European lovers of nature gaze longingly at the Wilderness Act, that has so far protected 110 million acres of land from significant human impacts. They also recoil from the ways in which some people still engage with protected lands: at the recent case, for example, in which a hunter in the Lolo National Forest in Montana repeatedly shot someone's pet malamute (and almost shot the owner) with a semi-

9. http://www.discovery.com/tv-shows/shark-week/videos/whale-attacked-by-megal-odon.htm.
10. Breeanna Hare, 9 August 2013, 'Discovery Channel defends dramatized shark special *Megalodon*', CNN, http://edition.cnn.com/2013/08/07/showbiz/tv/discovery-shark-week-megalodon/.
11. http://www.idahoforwildlife.com/2-content/39-salmon-predator-derby.

automatic assault rifle.[12] He was found to have broken no law, on the grounds that he believed the dog was a wolf, even though it was wearing an illuminated collar and its owner was screaming at him to stop.

As ranchers and hunters lobby—with some success—to remove the wolf's protections under the Endangered Species Act, other people are seeking to extend its range across the entire continent, by means of the world's most ambitious rewilding program. The four mega-linkages proposed by Dave Foreman and the Rewilding Institute would connect conservation areas from Baja, California, to southern Alaska, from central America to the Yukon, from the Everglades to the Canadian Maritimes and from Alaska to Labrador.[13] Their program seeks to reverse the fragmentation of habitats that has been driving local populations of many animals to extinction. It would create permeable landscapes, through which these animals could move once more. It hopes to restore the populations of large predators (such as wolves, bears, cougars, lynx, wolverines and jaguars) which would then begin to drive the dynamic ecological processes which permit so many other species to survive.

The plan is wildly ambitious, but it might not be as implausible as some people assume. As Foreman points out, even in Florida, where the human population has been rising rapidly and the politics are often difficult, the government, working with private landowners, has spent billions of dollars and added millions of acres to its conservation network, to reconnect fragmented ecosystems.[14] In the United States, perhaps more rapidly than anywhere else, farming is retreating from marginal and unproductive land; forests are returning and conservation easements and land trusts are proliferating.[15] The impossible dream is beginning to look credible, and to embolden similar movements throughout the US and in other parts of the world. I hope that this book will inspire you to support them.

12. John S. Adams, 10 December 2013, 'Pet malamute shot, killed by wolf hunter', *USA Today*, http://www.usatoday.com/story/news/nation/2013/12/10/pet-malamute-killed-by-wolf-hunter/3950523/.

13. Dave Foreman, 2004, *Rewilding North America: A Vision for Conservation in the 21st Century*, Island Press, Washington DC.

14. Ibid.

15. Adam Federman, 2013, 'Return of the Wild: Will humans make way for the greatest conservation experiment in centuries?', http://www.earthisland.org/journal/index.php/eij/article/return_of_the_wild/.

Acknowledgements

More than any other book I have written, *Feral* is a collaborative effort. While the words (except those quoted) are mine, and I have spent three years researching, writing and revising it, the ideas, structure and progress of this book have been developed with the help of many people, some of whom have given the project a great deal of their time and energy. I could not have written it without them.

I am very grateful to Helen Conford, Ketty Hughes, Antony Harwood, James Macdonald Lockhart, Ritchie Tassell, Alan Watson Featherstone, Clive Hambler, Mark Fisher, Miles King, Dafydd Morris-Jones, Delyth Morris-Jones, Paul Kingsnorth, Tomaž Hartmann, Jernej Stritih, Jony Easterby, Nick Garrison, Simon Fairlie, Morgan Parry, Peter Taylor, Bruce Heagerty, Jay Griffiths, Ralph Collard, Hannah Scrase, Michael Disney, Mick Green, Mark Lynas, Maria Padget, George Marshall, Annie Levy, Caitlin Shepherd, Estelle Bailey, Tammi Dallaston, Sharon Girardi, Mike Thrussell, Jean-Luc Solandt, Andy Warren, Jonathan Spencer, Jamie Lorimer, Adam Thorogood, David Hetherington, Paul Rose, Tobi Kellner, Steve Carver, Sophie Wynne-Jones, Ray Woods, Simon Drew, Miriam Quick, Leigh Caldwell, John Boardman, Meic Llewellyn, Guillaume Chapron, Staffan Widstrand, Kristjan Kaljund, Geoff Hill, Tom Edwards, Steve Forden, Paul Jepson, Joss Garman, Ann West, Clive Faulkner, Rod Aspinall, Liz Fleming-Williams, Grant Rowe, Ruth Davis, Elin Jones, Cath Midgley, Nick Fenwick, John Fish and Gary Momber.
The mistakes this book doubtless contains are all my own work.

I have changed the names of some of the places in mid-Wales in order to protect the wildlife I discuss from commercial exploitation.

If you find yourself in the area and ask Welsh speakers how to find the places I have mentioned, you are likely to be met with some very odd looks.

In the descriptive passages I have tended to use imperial measurements. When discussing scientific findings I have reverted to the metric system.

Introduction

It is an extraordinary thing for a foreigner to witness: one of the world's most sophisticated and beautiful nations being ransacked by barbarians. It is more extraordinary still to consider that these barbarians are not members of a foreign army, but of that nation's own elected government. The world has watched in astonishment as your liberal, cultured, decent country has been transformed into a thuggish petro-state. The oil curse which has blighted so many weaker nations has now struck in a place which seemed to epitomise solidity and sense.

This is not to say that there were no warnings in Canada's recent past. The nation has furnished the world with two of its most powerful environmental parables: one wholly bad, the other mostly good.

The story of the collapse of the North Atlantic cod fishery reads like a biography of the two horsemen of ecological destruction: greed and denial. The basis on which the stocks were managed was the opposite of the Precautionary Principle: the Providential Principle. This means that if there's even a one percent chance that our policy will not cause catastrophe, we'll take it. Foreigners and seals were blamed for the depletion of the fish, while the obvious contribution of the Canadian fleet and the Canadian government was overlooked. The fisheries science was rigged and, when it still produced the wrong answers, disregarded or denounced.[1] The government continued to sponsor bigger boats and new fish plants even as the stocks were crashing. A moratorium was imposed only after the fishery became commercially extinct: government and industry, after due consideration and debate, agreed that the non-existent fish should no longer be caught.

Even today, the best means of ensuring that stocks can recover and breed freely—declaring a large part (perhaps the majority) of the Grand Banks a permanent marine reserve in which no fishing takes place—has not happened. All over the world the evidence shows that such no-take zones greatly enhance the overall catch, even though less of the sea is available for fishing. But the Canadian government continues stoutly to defend the nation from the dark forces of science and reason.

The other great parable which still resonates with the rest of the world—the battle over Clayoquot Sound—began the same way: private companies were given the key to a magnificent ecosystem and told they could treat it as they wished. The forests would have followed the fishery to oblivion had it not been for a coalition of remarkable activists from the First Nations and beyond, who were prepared to lose their freedom—and possibly their lives—to prevent a great wound from being inflicted on the natural world. In 1994 they won, for a few years at least. Their courage in the face of police brutality and judicial repression inspired peaceful direct action movements all over the world.

So here are the two Canadas: one insatiable, blindly destructive, unmoved by beauty; the other brave, unselfish and far-sighted. There is no doubt about which of the two is now dominant. For Canada today is providing the world with a third parable: the remarkable, perhaps unprecedented story of a complex, diverse economy slipping down the development ladder towards dependence on a single primary resource, which happens to be the dirtiest commodity known to man.

The tar sands poisoned the politics first of Alberta then of the entire nation. Their story recapitulates that of the Grand Banks. To accommodate rapacious greed, science has been both co-opted and ignored, the Providential Principle has been widely deployed, laws have been redrafted and public life corrupted. The government's assault on behalf of the tar sands corporations on the common interests of all Canadians has licensed and empowered destructive tendencies throughout the nation.

Already the planned pipelines whose purpose is to transport the tar to new markets are carrying the toxic sludge of misinformation across

Canada. For example, the company hoping to build the Northern Gateway pipeline deleted from the animations it presented to the public one thousand square kilometres of islands, which lie across its tanker route down the Douglas Channel.[2] This had the effect of making the project look less threatening to the sensitive coastal ecosystems of British Columbia, and collisions less likely. It also strikes me as symbolic: if the natural world stands in the way, we will erase it.

Just as government and industry blamed and persecuted seals for the decline of cod in the North Atlantic, so have they blamed and persecuted other predators for the decline of woodland caribou. The Alberta Caribou Committee, which represents such defenders of the natural world as Petro-Canada, Shell, BP, ConocoPhillips, Koch Industries, TransCanada pipelines, Alberta-Pacific Forest Industries and the pulp company Daishowa Marubeni, came together to puzzle over the downfall of the species.[3] As there could not possibly be a link to the fragmentation of its habitat by seismic lines, pipelines, roads, oil platforms, timber cutting and the transformation of pristine forest into wasteland, the cause was at first mysterious.

But, after taking expert advice from one another, the committee members managed to solve the mystery. The problem was, of course, wolves. Although they have lived with caribou for thousands of years, and though caribou seldom feature in their diet,[4] wolves have suddenly become an urgent threat to the survival of the species, just as seals suddenly became cod's nemesis in the 1980s. The committee explained the nature of the problem to the government, which has responded by intensifying its poisoning and shooting of wolves, in order, of course, to protect the natural world.

But the resurrection of Grand Banks politics has also aroused the spirit of Clayoquot Sound. I see the emergence of the Idle No More movement as one of the most inspiring recent developments anywhere on earth. It demonstrates that the other Canada, though brutally trampled, has not died. The direct actions by the First Nations peoples who lead this movement, in defense of both the living planet and their own patrimony, remind the rest of the world that the Canadian government does not represent the will of all its people.

Even so, as the sheikhs of Saudi Alberta come to dominate federal politics, and as other provincial governments, harried by lobbyists working for destructive interests, feel licensed by the example of Edmonton and Ottawa to accede to their demands, the nature of Canada—in two senses—is changing with terrible speed. You can see this change manifested in the sharp decline of mountain caribou in British Columbia, caused primarily by logging;[5] in the salmon farms destroying the magnificent sockeye runs up the Fraser River; in the near-extinction of the greater sage-grouse; in the refusal to list severely threatened iconic species—such as polar bears, grizzlies, western wolverines, beluga whales and porbeagle sharks—under the Species at Risk Act;[6,7] in the failure to protect the boreal forests; in the auctioning of offshore oil rights in the Arctic, accompanied by the deregulation of oil spill response plans.[8]

The new Environmental Assessment Act and the gutting of the old Navigable Waters Protection Act suggest that this festival of destruction has only just begun. For those who appreciate natural beauty and understand ecosystem processes, it must feel like living in a country under enemy occupation. It must also be intensely embarrassing. Canada is becoming a pariah state, whose name now invokes images formerly associated with countries like Nigeria and Congo. Canadian friends joke that they stitch US flags onto their rucksacks when they go abroad.

So it feels odd, publishing a book about rewilding in a nation undergoing a rapid dewilding. But I hope that there are several respects in which it can be found relevant in Canada. The first is that it seeks to explain fascinating new findings in the science of ecology, which show that you cannot safely disaggregate an ecosystem. The loss of one species often has severe consequences for species and systems to which it appears at first to be unconnected. Killing predators, such as wolves and seals, can have paradoxical impacts, severely damaging the prey species and ecosystems that the culling claims to protect.

The next is that *Feral* provides a warning of what Canada's destination may be. With astonishing speed, in many places your complex and fascinating ecosystems are being reduced to near-deserts of the kind with which we are familiar in Europe. In the United Kingdom we have all but forgotten what we once had, and see our bare hills and empty niches as natural. Some of us find ourselves afflicted by an ill-defined longing, which I have come to understand as ecological boredom.

But perhaps most importantly, as I kept discovering over twenty-eight years of activism and campaigning journalism, sustaining the morale of people engaged in any political struggle requires a positive vision. It is not enough to know what you are fighting against: you must also know what you are fighting for. An ounce of hope is a more powerful stimulant than a ton of despair. The positive environmentalism I develop in Feral is intended to create a vision of a better place, which we can keep in mind even as we seek to prevent our governments from engineering a worse one.

My proposals are not in any sense a final answer, and they are likely to be developed in different ways in different places, but I will be happy if this book helps to stimulate new thinking about our place on the living planet and the ways in which we might engage with it. Nowhere, I believe, is in greater need of that than Canada.

I

Raucous Summer

I will arise and go now, for always night and day
I hear lake water lapping with low sounds by the shore;
While I stand on the roadway, or on the pavements grey,
I hear it in the deep heart's core

William Butler Yeats
The Lake Isle of Innisfree

Every time I lifted off a turf, the same thing appeared: a white comma, curled in the roots of the grass. I picked one up. It had a small ginger head and tiny legs. Its skin was stretched so tight that it seemed about to burst at the segments. In the tail I could see the indigo streak of its digestive tract. I guessed that it was the larva of a cockchafer, a bronze-backed beetle that swarms in early summer. I watched it twitching for a moment, then I put it in my mouth.

As soon as it broke on my tongue, two sensations hit me like bullets. The first was the taste. It was sweet, creamy, faintly smoky, like alpine butter. The second was the memory. I knew immediately why I had guessed it was good to eat. I stood in my garden, sleet drilling into the back of my neck, remembering.

It had taken me a moment, when I woke, to realize where I was. Above my head a blue tarpaulin rippled and snapped in the breeze. I could hear the pumps working, so I must have overslept. I swung my legs over the edge of the hammock and sat blinking in the bright light, gazing across the devastated land. The men were already up to their waists in water, spraying the gravel banks with high-pressure hoses. There had been some shootings in the night, but I could not see any bodies.

The images of the past few weeks crowded my mind. I remembered Zé, the serial killer who owned the airstrip at Macarão, taking his gunmen into the bar to liven things up, and the man who had been carried out with a hole the size of an apple in his chest. I thought of João, a *mestizo* from the north-east of Brazil, who had spent ten years crossing the Amazon on foot, walking as far as the mines in Peru and Bolivia, before cutting through the forests for another 2,000 miles to come here. 'I have killed only three men in my life,' he told me, 'and all the deaths were necessary. But I would kill that many again if I stayed here for a month.'

I recalled the man who had shown me the strange swelling on his calf. When I looked closely I saw that the flesh was writhing with long yellow maggots. I remembered the Professor, with his neat black beard, gold-rimmed spectacles and intense, ascetic manner, the cynical genius who managed the biggest claim for its scarcely literate owner. Before he came here he had, he said, been Director of the University of Rondônia.

But above all I thought of the man the other miners called Papillon. Blond, muscular, with an Asterix moustache, he towered over the small dark people who had been driven here by poverty and land-theft. He was one of the few, barring the bosses, the traders, the pimps and the owners of the airstrips, who had come to this hell through choice. Before he joined the goldrush the Frenchman had worked as an agricultural technician in the south of Brazil. Now, having found nothing, he was trapped in the forests of Roraima hundreds of miles from the nearest town, as destitute as the others. Here was a man who had leapt over the edge, who had abandoned comfort and certainty for a life of violent insecurity. His chances of coming out alive, solvent and healthy were slight. But I was not convinced that he had made the wrong choice.

I cleaned my teeth, picked up my notebook, then stepped out over the mud and gravel. The temperature was rising and in the surrounding forest the racket of yelps and whistles and trills was dying away. It was now three weeks since Barbara, the Canadian woman with whom I was working, had found a way through the police cordon at Boa Vista airport, and had shoved us, unrecorded, onto a flight to the mines. It felt like months. We had watched the miners tearing out the veins of the forest: the river valleys whose sediments were paved with gold. We had seen evidence of the one-sided war some of them were waging against the local Yanomami people, and the physical and cultural collapse of the communities they had invaded. We had heard the gunfire that came from the woods every night, as bandits waylaid the miners, thieves were executed, or men who had struck lucky fought over the gold they had found. In the six months since the main rush began here, 1,700 of the 40,000 miners had been shot dead. Fifteen per cent of the Yanomami had died of disease.

Now, because of the international scandal the invasion had caused, the new Brazilian government was clearing the mines, and moving the miners into enclaves in other parts of the Yanomami's land. From there, they knew, they could re-invade their old claims as soon as the rest of the world lost interest. The federal police had cut the supply lines: no planes had landed on the dirt airstrips for several days. The miners were using the last of their diesel and preparing to move. The police were supposed to have arrived the previous day, to confiscate weapons in advance of the expulsions, and the men had spent the morning moving in and out of the forest, burying their guns in plastic sheeting. I had stayed to watch, but the police had not come. Barbara had – Jesus, where the hell was Barbara?

She had set off yesterday to find a Yanomami village in the mountains and said she would be back that night. But no one had seen her. I cast around, through the shanties and bars the miners had erected, among the groups of men in the bottom of the pits, without success. I found my friend Paulo, a mechanic who had defended the indigenous people in arguments with the other miners, and we struck up the valley to look for her. The river ran orange and dead, choked by the forest clay disturbed by the mines. Around it, the valley was a wasteland of pits, spoil heaps and toppled trees. The miners who worked a stake called Junior Blefé told us that Barbara had passed through the previous day but had not returned. A man with a drinker's face and a black eye knew how to find the village and agreed to guide us. We set off, running, into the mountains.

Soon after we entered the darkness of the forest we began to find the prints of Barbara's plimsolls, a day old, overlain by the naked tracks of the Yanomami. I kept my eyes on the ground, but every so

3

often Paulo would stop and shout. 'Look at that water, look at those trees: so beautiful, isn't that beautiful?' I would stand and gaze for a moment, and see trees weighed down above clear water by moss and epiphytes, damselflies pausing in spots of light.

We ran on, following Barbara's footprints, slipping on the clay path. By midday we started to climb steeply; my breath came as if drawn through a sheet. Soon I saw light ahead of us: we were reaching the top of a mountain. From its crest we saw women on the far side of the valley, dressed only in loincloths, moving through banana groves, carrying baskets of fruit. Hills stepped away into silence, forested, undisturbed. We remained hidden among the trees for a few minutes, then we walked down to the lap of the valley and up into the gardens, calling out in Portuguese that we were friends. They stood still and watched us come close. I put out my hands and they shook them with shy grins.

'White woman,' I said. 'Have you seen the white woman?' I mimed Barbara's height and long hair.

They laughed and pointed up the slope behind them, into the forest. We began to run again, over the mountain and down into the next valley. We stumbled, exhausted, along the valley floor, tripping on roots, blundering into trees. We turned a corner of the path and stopped.

In the glade beside a stream a crowd of people sat or knelt, the honey of their skins cooled by the stained-glass light of the forest. The women wore feathers in their ears, the painted spots and stripes of wildcats; and jaguar's whiskers: stems of dried grass piercing their noses and cheeks. In the middle of the circle, radiant as a flower in the green dark of the forest, was Barbara.

She turned and smiled. 'Glad you could make it.'

The young Yanomami people led us along the path until we came to their *malocas*: round communal houses thatched almost to the ground with palm leaves. I took off my shirt and shoes – everyone else was nearly naked – and sat down. Children clustered around me, grinning and giggling, hiding their faces when I looked at them. They tugged at the hairs in my armpits: the Yanomami do not possess them. Someone gave me a plug of green leaves, and when I pushed it under my lip and sucked I forgot that I was hungry.

A young man came through the crowd and gestured that I was to help them build an extension to the communal *maloca*: they wanted me to climb to the top of the roof and tie on a tarpaulin they had been given by the miners. I stayed on the roof for a couple of hours, mending holes under his direction. When I came down I asked Barbara why he was so bossy.

'He's the chief,' she said.

'But he's only eighteen.'

She looked around. 'All the older men are dying or dead.'

In the living space of the *maloca*, the hammocks were filled with the sick. As I sat beside a feverish boy, two old women broke through the screen of banana leaves, shuffling on their haunches, roaring and sweeping sticks across the ground, their eyes screwed shut. I was hit on the ankles before I could get out of the way. The women stamped around the hammock, screaming, beating the air with their sticks.

The roaring continued for most of the day. I was later told that female faith healers were almost unknown among the Yanomami: only the absence of men could account for it. The old women led me to the hammock of a teenaged girl and showed me what I must do. I stamped and shouted, sweeping my arms through the air, scooping something from the surface of her body and pushing it away from the *maloca*. Urged on by the two women, I danced and yelled faster and louder, stamping and leaping over the hammock, until I almost fainted and fell into the arms of the healers.

When I had recovered and washed in the stream, the women brought me food laid out on a banana leaf: baked plantains, toad-stools and beetle grubs, foetally curled, still writhing. My hand hovered over the leaf. 'Go on,' they gestured. I picked up a grub and opened my mouth.

I leant on my spade, staring at the ground. On that raw December day soon after I had arrived in Wales, I was struck by the smallness of this life. Somehow – I am not quite sure how it happened – I had found myself living a life in which loading the dishwasher presented an interesting challenge.

The invasion of Roraima, which I had witnessed almost twenty years before, represents everything I hate. The miners, many of whom had been expelled from their own lands in the north-east of Brazil by businessmen and corrupt officials, were driven to the mines by poverty and desperation. But those who had organized it, who had the capital to build the airstrips and buy the machinery, were driven to kill and destroy by greed. Had the government of Brazil not changed, had the miners not, after several more months of procrastination, been expelled from the Yanomami's land, the tribe would have gone the same way as most of those in the Americas: to extinction. The old government knew this. Genocide was not its intention: simply an unavoidable, and unregretted, consequence of its policy.

And yet, even while I stayed in the goldmines and experienced the horrors of the invasion, I was drawn to what I hated. The mines exploded the metaphors by which we live. In the rich nations we trade in ciphers for gold, and seek them through specializations so extreme that we are in danger of losing many of our faculties. In the mines gold was gold, and the men got their hands dirty in all respects. Conflicts were resolved not through legal instruments or on the sofas of television studios, but by shoot-outs in the forest. It was rawer, wilder, more engaging than the life I had led; and the life I would lead thereafter.

J. G. Ballard reminded us that 'the suburbs dream of violence. Asleep in their drowsy villas, sheltered by benevolent shopping malls, they wait patiently for the nightmares that will wake them into a more passionate world.'[1] We still possess the fear, the courage, the aggression which evolved to see us through our quests and crises, and we still feel the need to exercise them. But our sublimated lives oblige us to invent challenges to replace the horrors of which we have been deprived. We find ourselves hedged by the consequences of our nature, living meekly for fear of provoking or damaging others. 'Thus conscience does make cowards of us all.'[2]

Much of the social history of the past two centuries consists of the discovery, often grudging, that other people, whatever their language, colour, religion or culture, have similar needs and desires to ours. As mass communication has enabled those whose rights we formerly disregarded to speak for themselves, to explain the impacts on their lives of the decisions we make, we become increasingly constrained by a necessary regard for others. Just as potently, we now know that little

we do is without environmental consequence. The amplification of
our lives by technology grants us a power over the natural world
which we can no longer afford to use. In everything we do we must
now be mindful of the lives of others, cautious, constrained, meticu-
lous. We may no longer live as if there were no tomorrow.

There are powerful and growing movements in many nations of
people who refuse to accept these constraints. They rebel against
taxes, health and safety laws, the regulation of business, restrictions
on smoking, speeding and guns, above all against environmental limits. Like the people who promoted the invasion of the Yanomami's
lands, they kick against the prohibitive decencies we owe to others.
They insist that they may swing their fists regardless of whose nose is
in the way, almost as if it were a human right.

I have no desire to join these people. I accept the need for limitations, for a life of restraint and sublimation. But I realized, on that
grey day in Wales, that I could not continue to live as I had done. I
could not continue just sitting and writing, looking after my daughter
and my house, running merely to stay fit, pursuing only what could
not be seen, watching the seasons cycling past without ever quite
belonging to them. I had offered too little to that life, the life of the
spirit,

> Which is not to be found in our obituaries
> Or in memories draped by the beneficent spider
> Or under seals broken by the lean solicitor
> In our empty rooms[3]

I was, I believed, ecologically bored.

I do not romanticize evolutionary time. I have already lived beyond
the lifespan of most hunter-gatherers. Without farming, sanitation,
vaccination, antibiotics, surgery and optometry I would be dead by
now. The outcome of mortal combat between me, myopically stumbling around with a stone-tipped spear, and an enraged giant aurochs
is not hard to predict.

The study of past ecosystems shows us that whenever people broke
into new lands, however rudimentary their technology and small their
numbers, they soon destroyed much of the wildlife – especially the
larger animals – that lived there. There was no state of grace, no

7

golden age in which people lived in harmony with nature. Neither do I wish to return to the hallows and gallows of the civilizations we have left behind.

Nor was it authenticity I sought: I do not find that a useful or intelligible concept. Even if it exists, it is by definition impossible to reach through striving. I wanted only to satisfy my craving for a richer, rawer life than I had recently lived. Yet somehow I had to reconcile this urge with the life I could not abandon: bringing up my child, paying my mortgage, respecting the rights and needs of other people, restraining myself from damaging the natural world. It was only when I stumbled across an unfamiliar word that I began to understand what I was looking for.

So young a word, yet so many meanings! By the time 'rewilding' entered the dictionary, in 2011,[4] it was already hotly contested. When it was first formulated, it meant releasing captive animals into the wild. Soon the definition expanded to describe the reintroduction of animal and plant species to habitats from which they had been excised. Some people began using it to mean the rehabilitation not just of particular species, but of entire ecosystems: a restoration of wilderness. Anarcho-primitivists then applied the word to human life, proposing a wilding of people and their cultures. The two definitions of interest to me, however, differ slightly from all of these.

The rewilding of natural ecosystems that fascinates me is not an attempt to restore them to any prior state, but to permit ecological processes to resume. In countries such as my own, the conservation movement, while well intentioned, has sought to freeze living systems in time. It attempts to prevent animals and plants from either leaving or – if they do not live there already – entering. It seeks to manage nature as if tending a garden. Many of the ecosystems, such as heath and moorland, blanket bog and rough grass, that it tries to preserve are dominated by the low, scrubby vegetation which remains after forests have been repeatedly cleared and burnt. This vegetation is cherished by wildlife groups, which prevent it from reverting to woodland through intensive grazing by sheep, cattle and horses. It is as if conservationists in the Amazon had decided to protect the cattle ranches, rather than the rainforest.

Rewilding recognizes that nature consists not just of a collection of species but also of their ever-shifting relationships with each other and with the physical environment. It understands that to keep an ecosystem in a state of arrested development, to preserve it as if it were a jar of pickles, is to protect something which bears little relationship to the natural world. This perspective has been influenced by some of the most arresting scientific developments of recent times.

Over the past few decades, ecologists have discovered the existence of widespread trophic cascades. These are processes caused by animals at the top of the food chain, which tumble all the way to the bottom. Predators and large herbivores can transform the places in which they live. In some cases they have changed not only the ecosystem but also the nature of the soil, the behaviour of rivers, the chemistry of the oceans and even the composition of the atmosphere. These findings suggest that the natural world is composed of even more fascinating and complex systems than we had imagined. They alter our understanding of how ecosystems function and present a radical challenge to some models of conservation. They make a powerful case for the reintroduction of large predators and other missing species.

While researching this book I have, with the help of the visionary forester Adam Thorogood, stumbled across an incendiary idea that seems to have been discussed nowhere but in a throwaway line in one scientific paper.[5] I hope it might prompt a reassessment of how our ecosystems function, and of the extent to which they are perceived as natural. There is, we believe, powerful circumstantial evidence suggesting that many of our familiar European trees and shrubs have evolved to resist attacks by elephants. The straight-tusked elephant, related to the species that still lives in Asia today, persisted in Europe until around 40,000 years ago,[6] a mere tick of evolution's clock. It was, most likely, hunted to extinction. If the evidence is as compelling as it seems, it suggests that this species dominated the temperate regions of Europe. Our ecosystems appear to be elephant-adapted.

Even so, I have no desire to try to re-create the landscapes or ecosystems that existed in the past, to reconstruct – as if that were possible – primordial wilderness. Rewilding, to me, is about resisting the urge to control nature and allowing it to find its own way. It involves reintroducing absent plants and animals (and in a few cases culling exotic species which cannot be contained by native wildlife), pulling down the fences, blocking the drainage ditches, but otherwise stepping back. At sea, it means excluding commercial fishing and other forms of exploitation. The ecosystems that result are best described not as wilderness, but as self-willed: governed not by human management but by their own processes.* Rewilding has no end points, no view about what a 'right' ecosystem or a 'right' assemblage of species looks like. It does not strive to produce a heath, a meadow, a rainforest, a kelp garden or a coral reef. It lets nature decide.

The ecosystems that will emerge, in our changed climates, on our depleted soils, will not be the same as those which prevailed in the past. The way they evolve cannot be predicted, which is one of the reasons why this project enthralls. While conservation often looks to the past, rewilding of this kind looks to the future.

The rewilding of both land and sea could produce ecosystems, even in such depleted regions as Britain and northern Europe, as profuse and captivating as those that people now travel halfway around the world to see. One of my hopes is that it makes magnificent wildlife accessible to everyone.

I mentioned that there are two definitions of rewilding that interest me. The second is the rewilding of human life. While some primitivists see a conflict between the civilized and the wild, the rewilding I envisage has nothing to do with shedding civilization. We can, I believe, enjoy the benefits of advanced technology while also enjoying, if we choose, a life richer in adventure and surprise. Rewilding is not about abandoning civilization but about enhancing it. It is to 'love not man the less, but Nature more'.[8]

The consequences of abandoning a sophisticated economy, supported by high crop yields, would be catastrophic. Before farming began in Britain, for example, these islands appear to have supported a maximum of 5,000 people.[9] Had they been evenly dispersed, each person would have occupied 54 square kilometres, an area slightly larger than the city of Southampton (which now houses 240,000 souls).[10] This, it seems, was as many people as hunting and

* This term was coined by Jay Hansford Vest.[7] It has been championed by Dr Mark Fisher, whose work has been influential in shaping this book.

gathering could sustain. (Even so, Mesolithic men and women severely reduced the numbers of large animals.) The fantasy entertained by some of the primitivists I have met, of returning to a hunter-gatherer economy, would first require the elimination of almost all human beings.

For the same reason I do not think that extensive rewilding should take place on productive land. It is better deployed in the places – especially in the uplands – in which production is so low that farming continues only as a result of the taxpayer's generosity. As essential services all over Europe (and in several other parts of the world) are cut through want of funds, farm subsidies in their current form surely cannot last much longer. Without them, it is hard to see how farming in these places can be sustained: for good or ill, it will gradually withdraw from the hills.

Some people see rewilding as a human retreat from nature; I see it as a re-involvement. I would like to see the reintroduction into the wild not only of wolves, lynx, wolverines, beavers, boar, moose, bison and – perhaps one day in the distant future – elephants and other species, but also of human beings. In other words, I see rewilding as an enhanced opportunity for people to engage with and delight in the natural world.

Feral also examines the lives we may no longer lead and the constraints – many of them necessary – that prevent us from exercising some of our neglected faculties. It explains how I have sought, within these constraints, to rewild my own life, to escape from ecological boredom. I am surely not alone in possessing an unmet need for a wilder life, and I suggest that this need might have caused a remarkable collective delusion, from which many thousands of people now suffer, that seems to be an almost perfect encapsulation of the desire for a fiercer, less predictable ecosystem.

If you are content with the scope of your life, if it is already as colourful and surprising as you might wish, if feeding the ducks is as close as you ever want to come to nature, this book is probably not for you. But if, like me, you sometimes feel that you are scratching at the walls of this life, hoping to find a way into a wider space beyond, then you may discover something here that resonates. I seek to challenge our perceptions of our place in the world, of its ecosystems and of the means by which we might connect with them.

In doing so, I hope to encourage a positive environmentalism. The treatment of the earth's living systems in the twentieth and early twenty-first centuries has been characterized by destruction and degradation. Environmentalists, in seeking to arrest this carnage, have been clear about what people should not do. We have argued that certain freedoms – to damage, to pollute, to waste – should be limited. While there are good reasons for these injunctions, we have offered little in return. We have urged only that people consume less, travel less, live not blithely but mindfully, don't tread on the grass. Without offering new freedoms for which to exchange the old ones, we are often seen as ascetics, killjoys and prigs. We know what we are against; now we must explain what we are for.

Using parts of Wales, Scotland, Slovenia, Poland, East Africa, North America and Brazil as its case studies of good and bad practice, *Feral* proposes an environmentalism which, without damaging the lives of others or the fabric of the biosphere, offers to expand rather than constrain the scope of people's lives. It offers new freedoms in exchange for those we have sought to restrict. It foresees large areas of self-willed land and sea, repopulated by the beasts now missing from these places, in which we may freely roam.

Perhaps most importantly, it offers hope. While rewilding should not become a substitute for protecting threatened places and species, the story it tells is that ecological change need not always proceed in the same direction. Environmentalism in the twentieth century foresaw a silent spring, in which the further degradation of the biosphere seemed inevitable. Rewilding offers the hope of a raucous summer, in which, in some parts of the world at least, destructive processes are thrown into reverse.

Nevertheless, like all visions, rewilding must be constantly questioned and challenged. It should happen only with the consent and enthusiasm of those who work on the land. It must never be used as an instrument of expropriation or dispossession. One of the chapters in this book describes some of the forced rewildings that have taken place around the world, and the human tragedies they have caused. Rewilding, paradoxically, should take place for the benefit of people, to enhance the world in which we live, and not for the sake of an abstraction we call Nature. *Disagree*

Researching this book has been a great adventure: this is the most bewitching topic I have ever explored. It has taken me to wild places, brought me into contact with wild life and wild people. It has exposed me to some of the most riveting findings – in the fields of biology, archaeology, history and geography – I have yet encountered. It has wrought deep changes in my own life. At times investigating these issues has felt like stepping through the back of the wardrobe. This story begins slowly, with my efforts to engage more fully with the ecosystems on my doorstep, to discover in them something of the untamed spirit I would like to resurrect. If you would care to push past the coats, you can join me there.

2

The Wild Hunt

I must go down to the seas again, for the call of the running tide
Is a wild call and a clear call that may not be denied

John Masefield
Sea Fever

On the riverbank, beside the old railway bridge, I loaded my boat. I tied on a spool I had made from hazel poles, wound with orange twine and a team of tinsel lures. I lashed a bottle of water and a wooden club to the cleats on either side of my seat, and attached the paddle to the boat with a leash: anything not tied down was likely to be lost. In the pockets of my lifejacket were spare lures, swivels and weights, a chocolate bar, a knife and – in case I was stung – a cigarette lighter.

I stepped into the brown water. It filled my diving boots, soaking into my socks. It would keep my feet warm all day. I pushed the boat into deeper water then swung myself into it and set off downstream. Two sandpipers dipped and swooped along the bank. A family of swans bow-waved up the river, struggling against the current. Soon I reached the fast sparkling water in the shallows beyond the first meander. It rose in plumes over the rocks and raced between them, breaking into manes of spray. I sped through the rapids, bouncing off the water cushions on the boulders, feeling alive and free. Then the river reached the beach and spilled in a shallow fan across it. I found a channel just deep enough to carry me, and slid down into the first wave, which swamped the kayak then let me pass. The other breakers alternately sluiced over the prow or lifted the boat to smack it down with a great shudder onto the water. I paddled hard, submerging, rising, collapsing into the troughs, pushing through the breaking waves into the rolling waters beyond.

I turned once, memorized the marks on the shore, then set out to sea. There was a moderate, irregular swell with a few white horses. The waves had the knapped faces of flints; their chipped crests spangled with sunlight. Ahead of me a fulmar glided down to the surface, half-wheeled then soared away.

I let the line out, lodged the spool beside my foot and passed the twine across my leg, just below the knee. As I paddled, I could feel the weight tripping across the rocks of the reef. Occasionally the line would drag, and I pulled it in to find clumps of crusty pink seaweed attached to the hooks, or leathery ropes of ribbon weed, sometimes twelve feet long. Half a mile from the land I crossed a band of lilac jellyfish. They could almost have been oilspots, a faint, two-dimensional bleaching of the water, but occasionally the wind would lift them, and they roiled, fat and rubbery, through the surface. They poured under the boat in their thousands. Some carried orange nematocysts on their tentacles. Seedy, segmented, the jellyfish looked like burst figs.

On the far side of the reef a crabber made his lonely rounds, hauling up his pots, rebaiting them, threading them back down the line as his boat chugged slowly between the buoys. I could smell the bait and diesel across half a mile of sea. He headed back to shore and I was alone.

Towards the edge of the reef the swell rose. The line felt its way through the sea like an extension of my senses, an antenna attached to my skin, twitching and trembling. From time to time the spool jumped up and the twine snapped taut across my knee, but when I stopped and pulled I felt only the weight dropping back as the wave that had lifted the line passed by. I was now a mile or more from the shore, but I had not yet found what I was looking for. Every time I encountered it, it seemed to be a little further from land than before.

A mile beyond the reef a gannet skimmed past me. It rose a few yards into the air, folded its wings and fell like a dart into the water, raising a plume of spray. It sat in the surface swallowing what it had caught, flew on then dived again. I gave chase, but still the line throbbed limply through the water. The sky had clouded over, the wind had stiffened, and now the rain began to spatter. The sea felt like a half-set jelly.

I paddled west for three hours, straight out to sea. The land became an olive smear, the seaside town to the south a faint pale line. The waves were rising and the rain pelted into my face like birdshot. I had travelled six or seven miles from the shore, further than I had been before. Yet still I had not found the place.

On the horizon, I saw a flock of dark birds. Convinced that they had found the fish, I raised my pace to ramming speed. They disappeared, then appeared again, whirling a few feet above the waves. As I came closer I saw that they were shearwaters, about fifty of them, rising, turning, then landing on the sea again. A knot of birds peeled off from the flock and circled me. Their black velvet wings almost brushed the waves. They were so close that I could see the glints in their eyes. They were not feeding – just looking. The faint sense of loneliness that had crept up on me as I headed away from land dispersed.

The birds settled on the water again and I stopped a short distance away. There was no sound except the sloshing of the waves and the wind, whistling high and very faint, through the shock cords on the boat. The birds were silent.

Every time I go to sea I seek this place, a place in which I feel a kind of peace I have never found on land. Others discover it on mountains, in deserts or by the methodical clearing of their minds through meditation. But my place was here; a here that was always different but always felt the same; a here that seemed to move further from the shore with every journey. The salt was encrusted on the back of my hands, my fingers were scored and shrivelled. The wind ravelled through my mind, the water rocked me. Nothing existed except the sea, the birds, the breeze. My mind blew empty.

I put down my paddle and watched the birds. They trod water, preserving the distance between us. Squalls of rain drummed against my forehead. The waves, higher now, lifted the bows and swung the kayak round: I had to pick up the paddle and occasionally turn the boat to face the wind. The drops raised little spines on the face of the waves. Here was my shrine, the place of safety in which the water cradled me, in which I freed myself from knowing.

After a while I began to move south, parallel to the distant shore. I travelled for about a mile, then stopped and allowed the wind to carry me. I might have drifted all the way to land, but I began to feel cold, so I started to paddle again. I was now so tired that, even with the wind behind me, the sea felt lumpy and stiff.

About three miles from the coast I passed two brown guillemots, dipping their beaks in the water, occasionally standing up to flutter their wings. As I paddled past them, they held their heads in the air, watching me from the corners of their eyes, but not leaving the sea. Soon afterwards I felt a sharp, unmistakable tug against my knee. I yanked the line, then pulled it in, hand over hand. I could almost hear the electric twanging on the cord. As the tackle approached the boat it jinked about crazily. I saw a white flash far down in the green, and soon afterwards pulled the fish into the boat. It bounced around on the deck, then drummed on the plastic with rapid shivers. I broke its neck.

The mackerel's back was the same deep emerald as the water, slashed with black stripes, which swirled and broke across the head. The belly was white and taut, narrowing to a slim wrist and the crisply forked tail of a swift. Its eye was a disc of cold jet. My fellow predator, cold-blooded daemon, brother disciple of Orion.

After another mile I felt the lightest tap on the line. I picked it up and pulled, but there was nothing. I pulled again and it was almost wrenched from my hand. Whatever had tugged before had come back when it saw the lures rise. This felt different: heavier and less jagged. The white flash showed me that I had three fish – a full hand. I hauled them in, trying to hold the line clear as they landed on the boat and threw themselves about: a moment's inattention would leave me with a twenty-minute tangle. As soon as I had stowed them I turned the boat and paddled back to where I had hooked them. I circled the water but could not find a shoal.

I ate the chocolate and tramped on. The sun flickered for a moment and the sea turned to fresh-cast lead. Then the clouds closed and the rain came down again.

Half a mile from the coast I hit a small shoal and pulled in half a dozen mackerel. Then I found myself in a strand of jellyfish so dense that in places it scarcely seemed to contain water. They poured under my boat in a column just a yard wide, heading away from land. The mackerel came up sporadically, in twos and threes. A driftline perhaps, which could explain why the predators had clustered around this strip: the plankton, like the jellyfish, had been corralled by a gentle rip tide, and the bait fish had followed them.

I watched the moon jellies rolling over each other like bubbles in a lava lamp. At one point the procession broke. There were a few yards of clear water, then I was startled by a monstrous ghostly jelly, pale and hideous, leading the next battalion. It took me a moment to see that it was a white plastic bag, parachuted taut in the water, the jellyfish king whose subjects followed him out to sea.

I drifted with them, sawing the line up and down. When I paddled, the jellyfish bumped against the line, causing me to stop and test the signal, to see what manner of life was tapping out its message from the gloom. I searched in vain for a baitball.

As usual in such matters, there were as many opinions about why the mackerel had scarcely appeared this year as there were people to ask. A local fishmonger told me with great authority that a monstrous new ship was operating in the Irish Sea, fishing not with a net but with a vacuum tube that sucked up the mackerel and everything else that came its way, which it turned into fishmeal for use as fertilizer and animal feed. It had been licensed by the Environment Agency to catch 500 tonnes of mackerel a day, and had received a £13 million subsidy from the European Commission. I checked this story and soon discovered that the Environment Agency has no jurisdiction at sea, that vacuum tubes are used not for fishing but for sucking the catch out of the nets, that there is no such fishmeal operation in the Irish Sea and that no boat is licensed to take such a tonnage. Otherwise the explanation was impeccable.

Others blamed the dolphins which, they said, had come into the bay in greater numbers than usual this year (the records suggested otherwise), or the north-west winds that had predominated since the end of May and were alleged to have broken up the shoals. Some people pointed to the black landings by a group of crooked fishermen in Scotland (they took £63 million-worth of over-quota mackerel and herring[1]); others to the failure by the European Union, Norway, Iceland and the Faroes to decide how many fish each nation should take, now that the shoals were moving further north in the winter;[2] or to overfishing in the Cantabrian Sea by the Spanish fleet, which had recently netted almost twice the tonnage its quota permits.[3]

I have not been able to establish whether or not the fish which migrate into Cardigan Bay belong to the same populations as those being hammered in other waters. In any case, the mackerel which enter the bay, even in better years, when you can pull 100 or 200 into the boat in an hour or so, are the tattered remnants of what was once a mighty population. Within living memory, local fishermen say, the shoals were three miles long;[4] today you would be lucky to find one which stretches to a hundred yards. The European Union classifies the mackerel stock in the Irish Sea as being 'within safe biological limits',[5] but this says more about our reduced expectations of what a healthy population looks like than the state of the species.

There was another bump on the line and I pulled up a small brown fish. I hesitated before I swung it in. Brown fish, on this coast, are brought in carefully, in case they belong to the species which, for anglers, is the most dangerous animal in British waters.

I first snared one on my virgin voyage into Cardigan Bay. I had been catching mackerel, which dashed around wildly when I hooked them. But this thing stayed down and shook its head. I could feel the vibrations all the way up the line. I brought it to the surface and saw that it was about eighteen inches long, etiolated, mottled brown and white.

As I lifted it out of the water it started thrashing madly. I swung it towards my free hand, but just before I grabbed it, some ancient alarm, long buried in the basal ganglia, sounded. I dropped the fish on the boat and studied it as it rattled around the deck. I thought I knew every species in British waters, but I had never seen anything like this. Fins ran the length of its body, shimmering purple and green. It had a snake's stripes on its flanks, bug eyes on the top of its head and a huge, upturned mouth. Suddenly, from some long-forgotten book or poster, the name swam into my mind.

This was not a member of the lesser species, which hides in the sand at low tide, ruining the holidays of bare-footed children. It was a greater weever, which, I later read, could make grown men weep and rage with pain. Like the smaller species, it has three poisoned spines on its dorsal fin and one on each gill cover. The pain, if not quickly treated, can last for days. A local woman, fishing on a charter boat, sat on one that someone had landed on the deck and spent six weeks in a wheelchair. A man I met was unable to move his left hand for six months. Few people have been killed by weevers, but if you are stung in a kayak and have no means of treatment, you will not make your own way back to land. The pain and shock ensure that paddling is impossible.

I managed, after nearly falling out of the boat, to shake the creature off the hook. Since then I have always carried a club with me. Whenever I catch a weever, I draw it against the side of the kayak and hit it very hard. It has firm white flesh, which makes an excellent bouillabaisse or curry. In the Mediterranean the charter boats allow anglers to take all the fish they catch except the weevers, which the crews keep for themselves.

On some occasions in the previous season I had caught weevers in greater numbers than mackerel. I had never been stung on the boat, but one day, filleting the fish on the shore while my partner made a fire in the dunes, my hand slipped and I impaled my thumb on a spine. It felt as if I had put my thumb on a workbench, raised a hammer and hit it as hard as I could. I went rigid with pain, then felt a panic-inducing numbness spreading up my arm, across my shoulder and into my chest. But, even as my brain flooded with red light, the wheels began to spin. The cure for weever stings is hot water, applied as quickly as possible. There was no hot water on the beach. But it could not be the water that cured you, as skin is waterproof. It must be the heat. The poison must be heat-sensitive. It did not matter what the source of heat was. Where was heat? I cast around, my eyes flickering, and saw the smoke rising from the dunes.

I ran up the beach, crouching over my arm, jumped over the dunes and thrust my thumb into the flames. My partner stared at me as if I had gone mad. But the effect was remarkable. Within a minute the pain began to subside. I held my thumb so close to the fire that it almost scorched; the pain from the flames was less urgent than the pain from the venom. Soon my screaming nerves fell still. The numbness subsided, and within half an hour I felt as well as I had before I impaled myself.

But the fish I brought into the boat now was not a weever. It had a high square forehead, a delicate beaked mouth, damasked chestnut flanks shot with gold, and crimson fins like Spanish fans, flecked with turquoise. Under the throat were long bony fingers, which it used to probe the sediments for food. Seen from the front, the tub gurnard looked like a goose, its eyes set high on the sides of its beaked head. From the side, it was as pretty as an aquarium fish. I released it and it flicked back into the deep.

Now the waves were breaking on the shingle a few hundred yards from where I sat. Still trailing the line, my arms heavy, legs trembling with effort, I made my way north, towards the row of white breakers on the edge of the reef. I wound the cord back onto the spool, secured the hooks and stowed it. Soon afterwards I crossed the salt barrier. It was a neat white line of foam. On one side the water was green and clear; on the other it was brown and turbid: fresh water pouring from the river and fanning out into the sea. The change was as abrupt as the colouring on a diagram.

I wove through the breaking waves. They beat themselves against the boulders in the rivermouth. They flicked the back of the boat around, threatening to tip me broadside into the rocky surf. I caught the end of a large roller; it swung me round and smashed the prow down onto a rock. I back-paddled, skidded across the face of the next breaker, then found a passage between two waves. My paddle bit the water and I pushed myself into the rivermouth.

The whitewater in the river had been slowed by the rising tide, and I was able, clinging to the inside of the meanders, to make way against it. Small flatfish torpedoed away beneath the hull. After a few hundred yards, the riverbed rose and the force of the water gathered. I hauled at the paddle, but soon came to a standstill. I wedged the paddle between the rocks and slid out of the boat. But, unstrung by tiredness, I lost my footing, fell headfirst into the water and caught my ankle in the paddle's leash. The boat started to drift downstream, pulling me with it. I thrashed until I grabbed the leash. I freed myself just as my face was being dragged under the water, then dived down the river to catch the kayak. I turned it and began to wade back upstream, so tired that I could scarcely breast the river.

In the quiet waters beyond the railway bridge I pulled the stern onto the shore, and shook the boat to slide the fish in the hold down to the bow hatch. Their backs had turned a deep aquamarine and their bellies had taken on a pink iridescent flash. They glowed in the evening light.

I fetched a board and another knife from the car. I filleted one of the mackerel, exposing the clean, translucent bone, then pinned the tail of the fillet to the board with my penknife and skinned it with the other knife. The flesh tasted of raw steak. I filleted two more fish and ate them. I sat on the riverbank for a while, watching the mullet dimpling the surface and the crows landing momentarily on the rusty bridge then flapping away when they saw me. I gutted the remaining fish. It was not a great haul, but for the first time on the boat that summer I had caught more energy than I had used.

3

Foreshadowings

In this world's youth wise Nature did make haste,
Things ripen'd sooner, and did longer last.

John Donne
The Progress of the Soul

It began with a call from my friend Ritchie Tassell. 'There's something I want you to see. How soon can you get here?'

'I'm on the beach. One hour?'

'That'll do.'

I threw my wetsuit into the car and set off around the estuary. If Ritchie, who had seen almost everything, thought it was worth my while, it would be.

In the marshes beside the track, the sedge warblers churred and buzzed. Swallows dipped over the ditches and flickered above the heads of the sheep. The scent of bog myrtle, which – honey and camphor – put me in mind of the Victorians, rose on the still air. Ritchie had lent me a pair of binoculars. We waited.

'There he is!'

At that distance, to my inexpert eye, it could have been a buzzard or a black-backed gull. But as it flapped up the estuary, with a strangely awkward beat, I noticed two things. First, that something was swaying and planing beneath it. Secondly, that it was too dark for a gull, too white for a buzzard. It took me a moment.

'Jesus H Christ on a bike!'

'That's what I said. More or less.'

'I can't quite believe what I'm seeing.'

'He's been here for three days. If he settles it'll be the first time since the seventeenth century.'

The bird was heading towards us. About twenty yards before it reached the track, it turned and flapped slowly past in profile. It was carrying a large flatfish. After another one hundred yards or so it landed on a fencepost and started tearing at the fish.

Ritchie was, indirectly, responsible. He had reasoned that the ospreys which had been breeding in Scotland since 1954 would migrate along this coast on their way to and from Africa, pausing to refuel in the estuaries and lakes. He had also guessed that the young birds would be looking for territory. He found the tallest spruce tree on his side of the valley, roped himself up, cut off the top and built a wooden platform fifty feet from the ground. He covered it in twigs and splattered white paint over it to look like droppings: this, apparently, is the best means of persuading ospreys to move in.

Across the valley, from his cottage beside the estuary, a keen naturalist had watched these preparations. It was not long before he had persuaded the local wildlife trust to build a platform of its own; it planted a telegraph pole beside the railway track, and nailed a sheet of plywood across the top.

'It was a no-brainer,' said Ritchie. 'He could choose a nice little residence deep in the woods, in the top of a tree overlooking the estuary, or an exposed pole right next to the railway line. Of course the little sod chose the wildlife trust's effort. Not that I'm bitter or anything.'

I was only half listening. I was still struggling to take in what I had just seen. My heart pounded. I was filled with wild yearning: of the kind that used to afflict me when I woke from that perennial pre-adolescent dream of floating down the stairs, my feet a few inches above the carpet. I had felt it only once in recent years; in fact just a month before I saw the osprey.

Demonstrating – as I do about once a fortnight – a startling absence of the survival instinct with which other people are blessed, I had launched my kayak from the town beach at Pwlldiwaelod into a ten-foot swell. On the way through the waves the boat had back-flipped, somersaulting over me and dashing my head on the shingle. I was lucky not to have been knocked out. Needless to say, I tried again. This time, I broke through the waves and paddled out to sea. Now, after catching a few fish, I was returning to land. The tide had risen, and ugly, jumbled breakers were smashing on the seawall. Two hundred yards from the shore, I hesitated. Even from where I sat I could see that the waves were stained brown by the shingle they flung up. I could hear them crack and sough against the wall. Fear ran over my skin like cold water. I scanned the shore for a better way in, but saw nothing.

Behind me I heard a monstrous hiss: a freak wave was about to break over my head. I ducked and braced the paddle against the water. Nothing happened. I turned round. The rollers came in steadily: high, white-capped, but, at this distance from the shore, not yet threatening. Astonished, I swivelled round, desperately seeking an explanation. It rose from the water beside the boat: a hooked grey fin, scarred and pitted, whose tip skimmed just under the shaft of my paddle. I knew what it was, but the shock of it enhanced my rising fear and I nearly panicked. I glanced this way and that, almost believing that I was under attack.

Then a remarkable thing happened. From the stern I heard a different sound: a crash and a rush of water. I turned and a gigantic bull dolphin soared into the air and almost over my head. As he flew past, he fixed his eye on mine. We held each other's gaze until he walloped back into the water. I stared at the spot, willing him to resurface, but I did not see him again. I turned and faced the shore once more, now without fear. Instead I felt a heart-wrenching exhilaration that lent me, for a moment, clarity. I studied the seawall and noticed something I had not seen before: a distant slipway taking the force of the waves. In its lee were two or three yards of calmer water.

I cut across the waves until, fifty yards from the shore, I lined up with their strike and pointed the prow at the quieter patch. It reappeared every few seconds as a breaker fell back; then it was swept away in the next assault on the wall. Above the roar of the waves, I could hear the pebbles rattling against the battlements like grapeshot, as the sea sucked and sagged at the stonework. I dug the paddle in and charged the shore. I held back for a moment as a wave rolled past me, then flew at the gap. I jumped from the boat as it slid into the lee of the slipway, and clambered onto the concrete wedge, just before the kayak smashed against the seawall. The collision reduced my fishing rod to splinters. It is a stretch to say that the dolphin saved my life, but without that shift in focus I might have been flotsam.

Twice in one year I had heard the call – that high, wild note of exaltation – after a drought of sensation that had persisted since early adulthood; a drought I had come to accept as a condition of middle age, like the loss of the upper reaches of hearing.

That night, after a pint with Ritchie and a long vigil in the garden, watching the light fading from the sky and the first of the stars flashing into view above the mountains, I was struck by something that had not occurred to me before. Flatfish live on the seafloor. Ospreys catch their prey close to the surface. The facts did not marry.

As soon as I could get away the following week, I took the boat into the estuary. I was hoping to see the bird again, but also to try to discover what the fish were doing. I missed the osprey. But after an hour or two of poking around the margins of the sandbanks, my question was answered. I found a spot in which flounders had gathered in such numbers that they rested not on the sand but upon each other. They lay in no more than a foot of water and cruised over my bare feet, shuddering away in puffs of sand when I moved.

I spent that evening in the garage, rummaging in boxes and pushing aside paint tins, flowerpots, flints, fossils and packets of seed. Long after I had ceased to believe in its existence, I found it underneath the bottles I had dug up in an ancient rubbish dump when I was a child. It was a small slim package wrapped in yellowed newsprint, spotted with rust and oil. I read:

A reunião aconteceu na Secretar–
–plicou o comandante de Polícia Fe–
–ará, no próximo dia 11 de Junho, d–

As I unwrapped the package, the paper disintegrated in my hands and the precious object fell into my palm. It was the first time I had seen it since I had bought it in a market beside the River Solimões eighteen years before. Hand-forged, beautifully finished, it had cost me less than a pound.

In a friend's overgrown hedge I found a hazel pole that grew straight for ten feet. I whipped the weapon to the pole with fine cord, then ran a stone along the points. They scarcely needed it: the trident was still needle-sharp. The shank had been left square and rough to take the cord, but the points were round, polished and perfectly tapered. Each had four barbs, identically angled and chamfered. It had been forged for harpooning *arapaimas* – among the largest freshwater fish in the world – but I would make do with lesser prey.

Two weeks passed before I could return to the water. I paddled to where the fish had been. But in the shifting sands in the middle of the estuary, there is no 'where': no fixed point to which you can return. I tracked back and forth like a dog that had lost the scent, beached the boat, waded through the shallows, crossed the channels and circled the pools. I saw nothing except the silvery mullet, which swirled away as they felt the kayak approaching. The flounders had gone; the flat-fish forum was buried under a sandbar.

Now, three years after I had first seen the osprey, I decided to try again. There was a gentle hubbub on the beach: an ice-cream van, a handful of cars, some children wading and splashing in the narrow runnels trapped by the sandbars when the tide had pulled the plug. Beyond the cars I saw a wonderful sight. An ancient woman wearing iridescent ski goggles and a blanket over her knees was riding her electric wheelchair at full tilt. Sand spurted from the wheels. She skidded around in tight circles, jolted forward and fishtailed through the ruts left by the cars. Someone's heart was still beating.

I looked across the rivermouth. It was dead low tide. On the sea this would be called low-water slack, but in the estuary there is no slack: water runs in odd directions throughout the cycle. Two broad channels and a web of creeks, some connected, some blind, cut through a desert of sand. Across the water the sun fell on the pastel shades of Drefursennaidd. The boats at anchor in the lane beside the harbour looked as bright as bath toys. The weather curtain fell half-way up the estuary: the hills beyond were hidden behind silver sheets of rain. It is like this for much of the year: Drefursennaidd has half the rainfall of Llanaelwyd, ten miles inland; Llanaelwyd, in turn, has half the rainfall of Mwrllwch, five miles to the north.

I strapped the spear to the side of the boat, rigged up an anchor, bowlined a dry bag to one of the cleats beside the stern well, loaded the pockets of my life-jacket with a knife, a notepad, Polaroids and a spool of cord and dragged the kayak down to a ditch in which a trickle of water still ran.

In this rill it looked as if a battle were being fought. Sand gobies shot off in puffs of smoke like artillery shells. Baby flatfish raised trails of ack-ack fire as they scudded away, tails hitting the mud every few inches. Battalions of heavy armour trundled sideways, claws swivelling towards me. Soon the water was deep enough to lift the boat, and I set off upstream.

It was dead still. The water rippled away from the kayak, startling giant mullet at the edges of the channel. They furrowed round in semi-circles, then shot away in explosions of spray. Ringed plovers pattered along the shore with strange throaty warblings, then glided ahead of me on sickle wings. I could smell rotting seaweed and hear the strange music of the mudflats: the fizz and snap of millions of tiny creatures shifting in their burrows. On the sandbanks was the wreckage of stumps and branches brought down by the recent floods.

A knot in brick-red breeding plumage ran along the sand dipping its head, then took off with a long swooping whistle. A bumblebee trapped in the surface film broadcast frantic barcode ripples: sound made visible. I stopped paddling and drifted upriver, into the maze.

As I moved up the estuary I started tasting the water. Salt meant that I was travelling up a cul-de-sac, fresh or brackish that I was following a channel connected to the river. On most days it worked. But so much rain had fallen in the past week that the water everywhere tasted slightly fresh: the tides must have been pushing it back and forth. I know of no other way of navigating the labyrinth of channels. There are no visual clues: even when you leave the boat and stand on the banks you can see only the major cuts. The runnels, which are two or three feet lower than the domed surface of the sand, are invisible until you are almost on top of them.

I paddled blindly and soon came to a network of bayous, trenches scoured out by the currents, connected only by a thread of water. I slipped out of the boat and began to drag it up this trickle. Whenever I stepped into deeper water I felt shrimps battering against my feet. They moved like a film missing most of its frames: they appeared, disappeared, appeared again a few inches away, darting with flicks so rapid that they were impossible to follow. On the bank a cormorant dried its wings.

The water was warm and murky, the colour of weak tea. The sand had settled into a pattern of scooped ripples, each of which had trapped a pool of dark humus: the ridges formed pale crescents as regular as wallpaper. The stream soon became navigable again, but now it was flowing towards me. I pushed on up, tasting, paddling, peering over the side. Orange-legged shore crabs backed into the sand as the shadow of the boat passed over them. Fat cockles lay a little agape, the shells edged with a pink frill of flesh. Cockle. The word rolled about in my head: round, hinged, opening and closing like the creature it described.

A curlew crossed the tidal desert ahead of me, casting its sad loop-ing call across the water. Lost in the flats, I no longer had a sense of scale. Rounding a bend in the stream, I was amazed to see two people standing on a sandbank. As I approached they opened their wings and flapped away. Sheep moved along the distant edge of the saltmarsh in single file.

The bayous coalesced into a wide, shallow pan. Wading across it, I felt something flutter over my feet. I turned and saw a brown dia-mond fluking away. It stopped just a few yards from me and buried itself. I marked the spot in my mind, swiftly unstrapped the spear, removed the corks and left the boat to drift, then stalked across to where the fish had settled. It could not have moved: I would have seen puffs of mud hanging in the water. But it had vanished. I probed a couple of likely-looking mounds, but the spear just sank into the sand. The flounder had disappeared like a ghost passing through a wall. I cast around, imagining that I must have lost the mark, but I found no trace of it.

I anchored the boat, removed my life vest and cagoule and drew from the dry bag an item seldom seen on a kayak, a white business shirt. I had realized, a few days before, that most of the birds that feed on fish – gulls, gannets, shearwaters, guillemots, herons, ospreys – have white bellies, enabling them to disappear against the sky. I stalked up the channel with the spear over my shoulder, moving my big feet as quietly as I could. I must have cut an odd figure.

I soon disturbed a flatfish – too small to spear – and watched as it settled back into the mud. Now I understood what had happened before. Instead of making a hump on the sand, it curled itself around a ripple, perfectly mimicking not only the colour but also the shape of the riverbed. Even when I hovered right over it, it was impossible to see. It did not dart off until I almost trod on it.

By now I had crossed the weather curtain. The wind whipped the water and the rain pitted the surface: spotting fish became still harder. One or two fair-sized flounders darted off, but into deeper water where I could see nothing. I went back and fetched the boat. As I paddled upstream, I saw the great slurping mouths of mullet protruding from the water. I was tempted to fling the spear at them, but knew that it was useless. Soon the stream I was following petered out in a wilderness of sand and empty cockle shells. It would take at least an hour for the tide to connect it to a main channel. The weather was worsening, so I turned round.

The flow had changed again: I had travelled against it in both directions. I returned to where I had seen a broken lobster pot marooned in the middle of the flats; now the sea was lapping round it. The wind rose; I struggled against air and water. As the tide flowed past me I marvelled at its filing system. There were lanes of twigs half a mile long, strands of seaweed, then a drift packed with what at first I took to be dead shrimps. There were millions: I feared for a moment that there had been a plague or a poisoning. But when I scooped some up I saw that they were cast-off skins: perfect little suits of armour, with a gauntlet for every pleopod and palp. Nowhere did I see twigs in the shrimp lane or shrimp skins among the seaweed; the current had chosen a stream for each of them.

A week later I tried again, perhaps for the last time. I launched the boat at the head of the estuary. My plan was to intercept the flounders on their way out of the tidal creeks that fed into the rivermouth. Here the drowsy summer pastures met the scoured flats of the windfunnel. Sheltered by bluffs and embankments, cattle flicked their tails in the deep July meadows. Two dabchicks flipped underwater as I approached; a kingfisher blurred along the bank.

I found the mouth of a stream, hidden between walls of reed. I passed between the banks, cut off by the rustling screens from other sights and sounds. The reeds gave way to wild banks of bramble, hemp agrimony, knapweed and vetch. Where an oak had fallen across the water, I stowed my paddle, lay back in the boat and pulled myself under the branches. The water was so clear that I seemed to be drifting through air. I could see every speck and fibre on the bed. But not only did I spot no fish, I saw no life of any kind: no beetles, skaters, nymphs or shrimps. No dragonflies patrolled the banks, no caddis or mayflies danced over the water. Perhaps this stream had passed through old lead mines. Lead has been worked here since the Romans, and even mines abandoned many years ago produce effluvia so toxic that almost nothing survives in the water it contaminates. Two streams meet in a village close to where I live. One bustles with trout and bullheads, the other is dead. One day, a friend who lives in the village tells me, the ducks kept there strayed from their usual haunts in the living stream and dabbled for a while in the other. They were all found belly up.

I slid down the stream and back into the estuary. As I rounded the last bend in the river, the wind buffeted me. I could see, across miles of water, all the way to the sea. Here, within the fortress of cloud that guarded the hills, the land was ochre, olive, viridian. Beyond the weather curtain, in the coastal sunshine, the fields, brightened by fertilizer, seemed almost to fluoresce. At the mouth of the estuary, the dunes appeared to float free of their surroundings. Separated from the foreground by a shimmering silver line, they hovered like Laputa over the mudflats.

A flock of Canada geese that had been bobbing and craning their necks on the bank took off, leaving a mess of moulted feathers tumbling over the mud. Merganser fledglings pounded the water as they flapped after their mother, who ungallantly abandoned them and circled the estuary. The tide was now roaring out. As it met the wind, it rose into standing waves, in which the boat seemed to be glued to the water: I had to lean forward and place the paddle almost beside the bows to make any progress. I travelled up creeks so narrow that when I met an obstruction I had to reverse out. I rode down the banks of the main channel, peering into the water, and saw nothing but mud and broken branches.

Before long the current pulled me past the mudbanks and into the empty quarter, the wide tract of sand I had explored before. But this time, riding the main channel, I found myself colliding with geysers, thick with sand and dead leaves, that rose unexpectedly in the middle of the river, sometimes with such force that I felt them thump and lift the boat as I passed over. A buoy buried in this boiling water seemed to plough away upstream, like a great fishing float pulled by a shark.

I drifted past the bank of a sandbar, my spear raised, scanning the clear water in the margins. The ubiquitous, uncatchable mullet exploded away. I startled two large flounders, but both ribboned off before I had a chance to thrust the harpoon. A platoon of oystercatchers in black and white uniforms, wings clamped smartly to their sides, turned as one body and marched across the sand as I approached. I saw the reflection of the spear on the water, and I was struck by a thought that had not occurred to me before: I was restoring the kayak to its original function. Both the technology and the name have been – like anorak and parka – borrowed from Arctic peoples. Just as I stalked the edges of the sandbanks with my harpoon, they patrolled the margins of the ice-floes. Here, however, they would have starved.

Local people had told me that flounder once swarmed the estuary in such numbers that they would push wheelbarrows down to the water and impale the fish with garden forks until the barrows were full. But after my last attempt, too late, I heard that the crab boats had recently begun netting flounder just beyond the mouth of the estuary, to use as bait. They had more or less cleaned them out. This practice, if the story is true, is so wasteful and (given the quantities of dead fish, as well as heads and bones, that the fishing industry discards) so unnecessary that it seems we have hardly moved on from the days when the English colonists in North America prised giant lobsters out of rockpools to feed to their pigs.[1] The least we should expect, in these lean times, is that any fish caught should be eaten by people.

I left the boat on a sandbank and waded for a mile or so over the ridged and furrowed bed of an emptying channel. The water had cleared: now I could see the bottom when I stood waist-deep. I moonwalked over the riverbed, almost weightless. Small flatfish catapulted out of the sand.

As I stalked up the channel, my spear poised above the water, I felt as flexed and focused as a heron. Every cell seemed stretched, tuned like a string to the world through which I moved, straining for a note among the shifting harmonics of wind and water. My concentration intensified until I became hyper-aware, sensing each grain beneath my bare toes, every ripple round my waist, every movement, however infinitesimal, among the benthos. Suddenly I was gone.

It is hard to explain what happened. Perhaps it was the mesmeric repetition of the ripples in the sand, perhaps an escalating pitch of attention that thrust me through the barrier of the present, but I was at that moment transported by the thought – the knowledge – that I had done this before.

Except for the two forays I have mentioned already, I had not. I do not believe in reincarnation, or in the persistence of a soul after the death of the body. Yet I felt that I was walking through something I had done a thousand times, that I knew this work as surely as I knew my way home.

I had experienced a similar flush of feeling once before. Foraging for herbs and fungi in a wood in southern England, I had pushed through a screen of branches and seen, beside a small stream, a ginger-brown mound. It was a muntjac, one of the Chinese barking deer that have proliferated here since they were released by the Duke of Bedford in the early twentieth century. It must have died a few minutes before I arrived. Its eyes were bright, the body warm. There was no wound, no trace of blood. Its fangs, the great hooked canines with which the bucks fight each other and rip dogs apart, protruded past the lower jaw.

This was forage on a different scale from that I had set out to find, and I hesitated for a moment, surveying the sleek tube of its body, the small coralline antlers, the tiny hooves. Then I gathered up the ankles and heaved it onto my shoulders.* The deer wrapped around my neck and back as if it had been tailored for me; the weight seemed to settle perfectly across my joints. The effect was remarkable. As soon as I felt its warmth on my back, I wanted to roar. My skin flushed, my lungs filled with air. This, my body told me, was why I was here. This was what I was for. Civilization slid off as easily as a bathrobe.

---

* Picking up an animal that has died of natural causes and taking it home is a foolish thing to do: when I phoned a veterinary surgeon I know to ask if I could eat the deer, he told me to bury it.

I believe, though I have no means of showing that this proposition is true, that in both cases I was experiencing a genetic memory. Through the greater part of human existence, while we were still subject to natural selection, we were shaped by imperatives – the need to feed, defend and shelter ourselves, to reciprocate and work together, to breed and to care for our children – which ensured that certain suites of behaviour became instinctive. They could be suppressed by thought but, like the innate response which makes a pensioner vault over a five-foot wall just before a truck ploughs into him, they evolved to guide us, alongside the slower processes of the conscious mind (which is shaped by learning and experience). These genetic memories – these unconsidered urges – are printed onto our chromosomes, an irreducible component of our identity.

Some of these stereotyped responses – like the instinctive ways in which we care for our children – are still appropriate and necessary. Others – such as the instincts which once helped us to defend ourselves and our families from both predators and competing clans – can cause disaster, in densely populated, technologically amplified societies, when they are unleashed. We have had to learn techniques of containment, to press our roaring blood into quieter channels. Where these urges are familiar to us, experience has taught us how to suppress or redirect them. But this sensation was new. I could not assimilate it because – until I picked up the deer – I had been unaware of its existence. It was overwhelming, raw, feral. I did not have a place to put it; but I knew that it belonged to me as much as the tendons I use to curl my fingers.

On the Welsh shore of the Severn estuary, archaeologists working with farmyard slurry scrapers have swept away 8,000 years of mud to reveal a fossil saltmarsh platform so well preserved that, when you see photographs of the footprints they have found, you look beyond for the beasts and people that left them. The Goldcliff excavation tells the story of a world before ours, to which we still belong.[2]

Some of the prints, left in loose mud, are big and sloshy; others clean and crisp. You can see the pads of the toes and the mud that welled up between them: the marks look as fresh as if they had been made on this tide. In some places the people had slipped and skidded, the tracks show how their heels swung round, their toes splayed to retain their balance. One set of prints trails a small hunting party of teenaged boys. They pause, turn, change pace together. The layer of mud over which they run is pitted with the tracks of red deer.

Another set reveals a group of young children larking in the mud: running in circles, skidding, kicking. But elsewhere the children – our great-grandparents to the power of 300 – moved more systematically. Even those as young as four appeared to have been foraging. 'It may be difficult for us to understand,' the archaeologists tell us, that children this small were happily gathering food, 'because of the western world's predisposition to over-protect the young.'[3] The pattern of adult tracks suggests that they might have been hunting birds or emptying traps.

Cutting across or skirting round the human prints are others: red and roe deer and the monstrous puddled spoor of giant aurochs. Two trails are immediately recognizable: dog. But they are not. Mesolithic dogs were about the size of a collie. Where they were kept, the sites are cluttered with chewed bones. These prints are too large, and associated with neither human marks nor other such clues: the evidence suggests wolves.

But the tracks that made my skin prickle belonged not to the mammals which still howl and bellow through our nightmares, but to quite another creature. Splayed across the lesser impressions of herons, oystercatchers, gulls and terns were caltrop prints six inches across, cut in the fossil mud like masons' marks. The tracks show, the researchers tell us, that the beast which left them was 'a very common breeding bird in the Mesolithic estuary'. Cranes. When I read that, I sat back and closed my eyes. I could almost hear their cornet cries echoing over the flats, and see them drifting in their hundreds down to the marshes on cloaked wings, hanging like paragliders as they tilted to land.

These beasts – four feet high, eight feet between the wingtips, the highest flying birds on earth, cruising at 32,000 feet – which hang in the air as if suspended on strings and fill the sky with sound as crisp and ethereal as the realms through which they travel, which, with dagger bills and cockaded tails, throw back their heads and dance in the courting season, springing from the ground on extended wings and descending so slowly that they seem to be as light as air, once thronged the estuaries and wetlands. They lived in Britain in such numbers that, when George Neville became Archbishop of York in 1465, he served 204 of them at his inauguration feast.[4] This could help to explain why they became extinct here 400 years ago. But in 1979 they began to creep back. Birds migrating from the Continent established a small breeding colony in Norfolk, encouraging conservationists to try to reintroduce them elsewhere. In 2009, a group was released in the Somerset levels.[5] They will, their mentors hope, spread up the Severn valley into the quags and slobs of the rest of Britain. The findings at the Goldcliff dig augur well for the first phase of their expansion.

Among the tracks the archaeologists found the remains of Mesolithic meals. Here were the bones of red and roe deer and wild boar, charred and marked with stone axe cuts, and the colossal ribs and vertebrae of giant aurochs, one of which had been chipped by an arrow head or spear; a few otter and duck bones, charred hazelnut, cockle and crab shells. Two microliths – the small stone blades with which spears and arrows were tipped – have been oxidized by fire, which suggests that they were still lodged in meat that was cooked here. But overwhelmingly the remains are of fish: salmon, pouting, bass, mullet, flatfish and, above all, eels. The number and size suggest that the people trapped them here in shallow water, on moonlit stormy nights around the autumn equinox, when the eels began the migrations that would take them to the far side of the Atlantic. Three pointed stakes uncovered in a fossil channel could once have supported a set of basket traps.

I remember those movements from my own childhood: standing beside clear streams in Norfolk and the southern counties and watching a black chute of eels, which sometimes looked as densely twined as wickerwork, writhing its way downriver. Now you would be fortunate to see half a dozen in a day. The great caravan persisted from the Mesolithic until the 1980s, then collapsed.

Among the stone blades and grinding stones and adzes, the awls and scrapers made of bone, the antler mattocks scattered over the fossil marshes, were artefacts seldom found in sites of this age: tools made from wood. The excavators found a spatula, a wooden pin, a digging stick. But the one that intrigued me was a y-shaped stick, abraded, perhaps by sand, on the inside of the fork. The researchers believe that this might have been used to trap eels hiding in the sediments, pinning them down until they could be grabbed. I thought of those people stalking the channels with their prongs, walking slowly so as not to telegraph their movements through the water, their feet settling into the sand, scanning the bed for the faint trail of mucus or the serpentine mound that marked their quarry; raising the stick, adjusting for refraction, plunging it down. The eel whips and loops, snaking around the hand that seizes it. The fingers bite into the slimy flesh behind the gills, lift it out, thrash the tail against the pole to break the spinal column. The hand then pushes a stripped willow wand through the gills and out through the mouth, and slings the eel, with the rest of the prey, from the leather thong the hunter had tied around her waist.

Remnants in the mud suggest that these people camped on the salt-marsh platform in tipis. A structure nine feet across, with skins or reeds trussed over the poles, would have housed four people. They used the hearth at the centre to keep themselves warm and to roast or smoke their food. Exposed to the wind and rain of the Welsh coast in the bitter climate after the glaciers retreated, they must have been as tough as a lamb chop in a motorway service station.

We know little about British life in the Mesolithic: the near 6,000 years (between 11,600 and 6,000 years ago) after the retreat of the ice sheets, partly because much of the land over which those people roamed is now under water. At the end of the last glacial period, the sea level was 30 fathoms* lower than today's.[6] When the Mesolithic began, some 4,000 years before the camps discovered at Goldcliff were pitched, there was no Bristol Channel, no Cardigan Bay, no Liverpool Bay. Even Lundy Island, which marks the western end of the Bristol Channel, belonged to the mainland. But the sea rose with great speed. Evidence of human occupation at the Goldcliff site begins (about 7,800 years ago) when the sea first reaches it. By that time most of Cardigan Bay was under water, and the seas were still rising, about one and a half times as fast as they are today.

Like most coastal places, mid-Wales has its Atlantis myth, which might, though it was doubtless updated with the telling until it was

* 180 feet or 55 metres.

finally fixed in the written record, have originated in the drowning of settlements as the seas expanded after the Ice Age. The Welsh story tells of the Cantre'r Gwaelod – the Lowland Hundred – ruled by a chieftain called Gwyddno Garanhir. It was defended from the sea by a series of dykes. Gwyddno's nobles were in charge of maintaining the dykes and their gates and hatches. Among them was the notorious drunkard Seithenyn. He was on duty on the night of a terrible storm surge, with predictable consequences. The legend insists that the submerged bells of Cantre'r Gwaelod ring out when someone is in trouble at sea. I can testify that this story is untrue: I would have heard them often enough.

The evidence at Goldcliff suggests that the people who left their traces there hunted and foraged on the marshes only periodically, mostly in the summer and early autumn. Like the other predators, they followed the great herds of deer and aurochs, the sounders of boar and the boom-and-bust abundance of the rest of the natural system. They appear to have set up camp on the saltmarsh for a few weeks at a time, when the game filled the coastal forests and fish thronged the water. Emerging from the buried soil are great stumps and fallen oak trunks: some of which have no branches for forty feet. This suggests a closed canopy forest, rising from just above the high-tide mark. The mud contains the pollen of oak, birch, pine, hazel, elm, lime, alder, ash and willow. Along the shore were reedbeds, raised bogs and alder carr (swamp forest). Around the roots of the trees, the archaeologists found stores of hazelnuts, buried by Mesolithic red squirrels.

These people, they speculate, as well as hunting fish and game, would have eaten the roots and shoots of the reeds, the sweet gum oozing from the rushes, the seeds of grass and orache, barkbread from the birch trees, nuts, acorns, leaves and wild fruit. Evidence from other parts of Britain and Europe suggests that they are likely to have used dug-out canoes to hunt and gather in the estuary and to travel to hunting grounds further along the coast.

In late autumn they might have migrated to beaches where seals heaved themselves out of the water to breed: easy prey for anyone who could reach them before they flopped into the water. In winter, they moved inland, hunting migratory birds in the upper estuary and the beasts in the forests. The growth patterns of the cockleshells in the Mesolithic middens of north Wales suggest that they were picked in spring and early summer, when they were fattest: the Goldcliff people might have travelled down the estuary to find them. When the cockles were over, hunting groups moved into the mountains, following the deer migrating to the greening pastures above the treeline. Then, it seems, they moved back down to the shore to intercept the fish migrations. There might have been places to which they returned every year, but they had no home. They moved with their prey, scattering fragments of their lives as they went: stone tools on the mountaintops, heaped shells on the seashore, weapons in the woods, chipped bones, decorated pebbles, an occasional burial. In the fossil marshes at Lydstep in Pembrokeshire, archaeologists have found the skeleton of a wild boar in which two microliths were embedded: carrying the arrow or spear that had wounded it, it plunged into the swamp to die.[7]

I looked again at those footprints receding across the marsh and into time. I heard the noise of the children playing in the mud, saw the tense, grave faces of the hunting party, watched in my mind's eye the women and elders wading along the estuary with their spears and prongs, and I felt I knew better who I was; where I have come from; what I still am.

4

Elopement

My friend, blood shaking my heart
The awful daring of a moment's surrender
Which an age of prudence can never retract
By this, and this only, we have existed

T. S. Eliot
'The Waste Land'

I turned away, trying to disguise my delight. At last, and quite by accident, I had found something he was afraid of.

'George, please, I am asking you, do not touch that thing.'

'It's harmless.'

'No! Very very harmful. Very poison.'

He backed away, shaking his head. Six months had passed since I had first met him, six months in which nothing and no one had ruffled his smooth humour, in which his feats of daring had left me – though I prided myself on plunging headfirst into danger – feeling like a chicken. With a sense of cruel triumph, I put my hand into the bush.

'George, I am asking you . . .'

The chameleon swivelled a turret eye to study my hand and flushed faintly russet. I gently pushed a finger under one of its feet, and the pincer toes clamped round it. I lifted the rest of my hand under the creature. It clung on, and I slowly raised it out of the bush. It turned a pale brick colour.

Toronkei had backed off to five yards away. I could see the sweat starting up on his forehead. His lips were working, but no sound emerged.

'You see, harmless. It's a myth.'

He edged forward. This time *his* pride was piqued. The chameleon sat quietly on my hand, rotating its eyes. It wound its tail around my little finger.

'You can touch it if you want. It won't hurt you.'

Clutching his spear so tightly that his knuckles shone, Toronkei advanced towards me. His mouth hung open. Trembling with self-control, he stretched out a hand and pushed the tip of his finger forward until it touched the chameleon's flank. It reared up, opened its pink mouth and hissed. He leapt backwards, stumbled, almost fell. Now it was my turn to struggle to control myself. I turned away and returned the chameleon to the bush, desperately trying not to laugh. I pretended to watch it settle in for a moment while I rearranged my face, then turned back. Toronkei stared at me with what I chose to believe was new respect. It is more likely to have been a conviction that I had gone mad.

Setting off at dawn, we had already run and walked twenty miles, describing a wide loop across Kajiado District, in the northern part of the Maasai's territory. At midday we had stopped at his uncle's house for milk, and spent two hours sitting in the shade, talking and swatting away flies. Now, with fifteen miles to go, we were travelling home to Toronkei's *manyatta*. We stood on a low escarpment, looking across the plains, spotted with shrubs and thundercloud acacias, that rose, through sage to grey to blue, towards an invisible Kilimanjaro, shrouded, as it so often was, by cloud or the mere thickness of the sky. Wavering through the heat haze beneath us were herds of multi-coloured cattle, dun eland, impala.

As usual, Toronkei had outpaced me, but every so often he had stopped and pretended to scan the land to allow me to catch up; he was more protective of my feelings than I was of his. We had no particular objective, other than visiting his uncle; running over the savannahs was an end in itself. He and the other *moran* would push themselves to accomplish remarkable feats, such as driving their cattle 140 miles in three days, without eating, drinking or sleeping. Occasionally, though they were now severely punished if caught by the Kenyan police, they would raid cattle from the Kikuyu who lived in the surrounding lands, sometimes escaping under a storm of bullets. Talking to Toronkei and the other warriors, it had struck me that escaping under a storm of bullets was as much the purpose of the exercise as stealing the cattle. In crossing and recrossing their wide lands, the *moran* came to know them as well as we know our own suburbs.

I had followed Toronkei through the defining phase of his life. He had been circumcised six years before I had met him. During the operation he had had to sit calmly, without twitching or blinking. Those who succeeded were given cattle; those who flinched would be ostracized. The warriors trained themselves to overcome pain: Toronkei had a circular scar on each thigh, where he had pressed glowing embers into his flesh.

Now, at nineteen, he had begun the long round of the warriors' graduation ceremonies, at the end of which they would acquire the status of junior elders, and be permitted to marry and set up their own homes. I had watched him, across the course of months, dancing, carousing and travelling with the other *moran*. I had seen them catch a sacrificial ox by the horns and tail – it flung them across the *manyatta* until they overpowered it – force it to drink a gourd of beer, then suffocate it and drink its blood. I had witnessed the strong bonds of love between the warriors, but also seen how their knives appeared from under their cloaks as soon as an argument began.

They had – though I had not seen it – killed a lion, in the manner tradition prescribed: they cornered it, one of them caught it by the tail and the others sought to spear it to death. Nothing appeared to perturb the *moran* – except chameleons. Danger to them was a delicacy, to be sought out and savoured. They were volatile, passionate, impetuous, open to everything. Perhaps because, being nomadic, they mixed with many cultures, I found it easier to engage with them than with the indigenous people among whom I had worked in West Papua and Brazil. They accepted me in the same spirit as they accepted everything else that came their way; nothing was permitted to impede experience. Though I was eleven years his senior, Toronkei and I, in a way that had not been possible elsewhere, became friends.

A few weeks after we had run to his uncle's house, I returned to Toronkei's *manyatta*, to watch the last of the ceremonies. The *moran* were dancing slowly and sadly, with a gentle murmur like the wind in the trees. The years of wild adventure were coming to an end. As I watched, a young man strode up to the edge of the group carrying the long, loosely spiralling horn of a greater kudu antelope. He put his mouth to a hole in the horn and blew four loud blasts, so deep that I felt them vibrating through my body. Screaming and howling, the dancers scattered, knocking me over. Four or five warriors collapsed and lay on the ground, quivering and groaning. People tried to pull them to their feet, but they seemed to be unconscious. They growled, drooled and blew. Their heels drummed on the ground. The horn was blown only in the last days of graduation, and whenever they heard it the warriors were overwhelmed with grief.

I followed Toronkei into the graduation hut that his mother had built for him – a small wicker box rendered with cow dung – and crouched for a while beneath the low ceiling until my eyes adjusted to the darkness. When I could see, I noticed an unfamiliar woman sitting on the cowhide pallet. She was very dark, with strong eyebrows, a smooth, round forehead and a cool, almost mocking look. I introduced myself. She turned away with an oddly bashful smile. I looked at Toronkei, puzzled, and was surprised to see that he was laughing.

'This,' he said, 'is my wife.'

Three days before I arrived in the *manyatta*, he had run thirty miles to visit a friend. As he approached the friend's village, he met the girl walking up the track, and changed his plans. They spent the day together, and by nightfall he had persuaded her to elope with him. They waited until everyone in her village was asleep, then slipped out of the compound and ran. The dogs woke, and her brothers set off in pursuit. The two lovers darted through the scrub, but soon after midnight the brothers surrounded them. The girl refused to go home. She told her brothers that if they wanted to talk to her they would have to come to Toronkei's village. The brothers returned to their compound, and Toronkei and his fiancée reached his *manyatta* just before dawn.

Her father was furious, but there was little he could do: his daughter would not be dissuaded. Toronkei had opened negotiations: the father had demanded a bride price of five cows and 10,000 shillings. Toronkei's parents were trying to talk him down. The girl came from a rich family, and the deal would be tough.

Hearing this story, watching the proud, conspiratorial looks he exchanged with his bride, seeing the hero's treatment he now received from the other *moran*, I felt, not for the first time in my friendship with Toronkei, a spasm of jealousy. I sat in the hut drinking milk and greeting the procession of young men who came in to pay their respects to him, troubled by a sense of inadequacy. As I watched the warriors sitting hand in hand on the pallet, and the young woman looking tenderly at her husband, I was struck by a thought so clear and resonant that it was as if a bell had been rung beside my ear. Had I, as an embryo, been given a choice between my life and his – knowing that, whichever I accepted, I would adapt to it and make myself comfortable within it – I would have taken his.

Despite six rich years of adventure in the tropics, mine now looked like a small and shuffling life. I thought of what awaited me when, in a few months' time, I returned home. I had been planning to finish my book, find new work, rekindle old friendships, perhaps put down a deposit on a house. After two bouts of cerebral malaria, as my expenses mounted and my savings trickled away, as I tired of lice, mosquitoes, foul water and corrugated roads, it had seemed appealing. But now I thought of the conversations confined to the three Rs: renovation, recipes and resorts. I thought of railings and hoardings. I thought of walks in the English countryside, where people start shouting at you as soon as you stray from the footpath. I succumbed, not for the first time in my life, to an attack of the futilities.

In 1753, Benjamin Franklin, writing to the English botanist Peter Collinson, made the following complaint:

> When an Indian Child has been brought up among us, taught our language and habituated to our Customs, yet if he goes to see his relations and make one Indian Ramble with them, there is no perswading him ever to return, and that this is not natural to them merely as Indians, but as men, is plain from this, that when white persons of either sex have been taken prisoners young by the Indians, and lived a while among them, tho' ransomed by their Friends, and treated with all imaginable tenderness to prevail with them to stay among the English, yet in a Short time they become disgusted with our manner of life, and the care and pains that are necessary to support it, and take the first good Opportunity of escaping again into the Woods, from whence there is no reclaiming them.[1]

Elopement with indigenous peoples was seen by the colonial authorities as a major threat to their attempts to subjugate the New World. When, in 1612, young men started defecting from Jamestown, the first sustained English settlement in North America, the deputy governor, Thomas Dale, hunted them down. According to a contemporary account,

> Some he apointed to be hanged. Some burned. Some to be broken upon wheles, others to be staked and some to be shott to death.[2]

The severity of these sanctions hints at the strength of the attraction. Despite the penalties, Europeans continued to defect, or to remain with the indigenous peoples who had captured them in war, until the Native Americans had been so reduced and broken that there was no longer a life to be drawn to. In 1785, Hector de Crèvecoeur remarked upon the fierce determination of European children to stay with the Indian communities that had kidnapped them, when their parents came to collect them during periods of peace.

> ... those whose more advanced ages permitted them to recollect their fathers and mothers, absolutely refused to follow them, and ran to their adopted parents for protection against the effusions of love their unhappy real parents lavished on them! Incredible as this may appear, I have heard it asserted in a thousand instances, among persons of credit. In the village of ------, where I purpose to go, there lived, about fifteen years ago, an Englishman and a Swede ... They were grown to the age of men when they were taken; they happily escaped the great punishment of war captives, and were obliged to marry the Squaws who had saved their lives by adoption. By the force of habit, they became at last thoroughly naturalised to this wild course of life. While I was there, their friends sent them a considerable sum of money to ransom themselves with. The Indians, their old masters, gave them their choice ... They chose to remain; and the reasons they gave me would greatly surprise you: the most perfect freedom, the ease of living, the absence of those cares and corroding solicitudes which so often prevail with us ... thousands of Europeans are Indians, and we have no examples of even one of these Aborigines having from choice become Europeans![3]

The encounter between the Old and New Worlds was characterized by dispossession, oppression and massacre, but in some places there were periods of friendly engagement. As Crèvecoeur documents, Native Americans were sometimes given the opportunity to join European households as equals; and in many cases Europeans were able to join Native American communities on the same basis. It could be seen as a social experiment. In both instances, people had a choice between the relatively secure, but confined, settled and regulated life of the Europeans, and the mobile, free and uncertain life of the Native Americans. There was no mistaking the outcome. In every case, Crèvecoeur and Franklin tell us, the Europeans chose to stay with the Native Americans, and the Native Americans returned, at the first opportunity, to their own communities. This says more than is comfortable about our own lives.

So why did I not defect to Toronkei's community? It is a question that still troubles me.

I was, as I had kept discovering, too soft for his life. I could not quite keep up physically. More importantly, I could not cope with the uncertainty: with the dislocation of not knowing whether I would eat today or eat tomorrow, or still possess a living – or a life – in a month's time. The Maasai accepted wild fluctuations in their fortunes with equanimity. In one season, their cattle would darken the plains; in the next, drought struck and they had nothing. To know what comes next has been perhaps the dominant aim of materially complex societies. Yet, having achieved it, or almost achieved it, we have been rewarded with a new collection of unmet needs. We have privileged safety over experience; gained much in doing so, and lost much.

But, perhaps overwhelmingly, I was aware that the old life was over. The Kenyan government was breaking up the Maasai's lands. Powerful elders were seizing as much as they could lay hands on; now the others scrambled to grab something for themselves. The community was collapsing; there was no common land left on which *manyattas* could be built and ceremonies held. As the power structures changed, the age groups, around which the life of the Maasai had been constructed, became an anachronism. Toronkei's was the last generation of warriors that would graduate in his community. The people were beginning to settle down, to move to the cities, to lose the freedoms which distinguished them from us.

But even had these pressures not existed, the wild life of the *moran* would have become less viable. Lion hunts are now severely punished by the Kenyan authorities, as lions are becoming scarce. The principles of universalism are arriving slowly in Kenya, where politics still divide people on tribal lines. But I doubt that the Kikuyu have ever enjoyed having their cattle raided – and their warriors speared – by the Maasai. As groups other than our own are able to make their needs and their rights known to us, as we come to recognize their humanity, we can no longer subordinate their lives to our desires; no longer expand our world into theirs. The freedoms the Maasai enjoyed at the expense of others – thrilling as they were – are rightly being curtailed. Perhaps there is no remaining moral space for the exercise of physical courage. Wherever you might seek to swing your fist, someone's nose is in the way.

Though it is now almost universally admired, when Jez Butterworth's play *Jerusalem* began to be noticed it sharply divided its audience. At the end of the performance I watched, in the last week of its first, incandescent West End run, half the audience stood to applaud, the rest barged out with thunderous faces, snapping and muttering.

Johnny Byron, played mesmerically by Mark Rylance, is the last of the Mohicans. He is sensuous, feckless, promiscuous, wild and free. He is a charismatic but ignoble savage, living in a mobile home in the woods, mad, bad and dangerous to know, the last man in England still in touch with the old gods. His totemic creature – his avatar – is the giant he claims to have met and whom he insists he can rouse: the undiminished ancient being, free from regulation or social constraint, who no longer belongs to a world in which new estates crowd the woods and council officers in yellow jackets patrol with their clipboards.

'Grab your fill,' Byron tells us. 'No man was ever lain in his barrow wishing he'd loved one less woman. Don't listen to no one and nothing but what your own heart bids. Lie. Cheat. Steal. Fight to the death.'

He lives by this creed, the curse of officialdom, the bane of the tidy, sedentary people who hate and envy him, a drug-dealer, fighter, seducer, former daredevil, teller of tall tales, magnet for disaffected teenagers, scabby, piss-soaked, drunken prince of revelry, master of the last wild hunt. He is pitched against his childhood friend Wesley, now the landlord of the local pub (from which, of course, Johnny has been banned), who is ground down by the demands of the brewery, by health and safety regulations, by his humdrum, responsible life and the sanitized, pasteurized world he has created. '... fiddly bloody sachets, broken bloody towel dispensers, fucking stupid T-shirts. I come to bed when the last cunt's gone home. I lie there next to her and I can't breathe ... Number one, work all your life. Number two, be nice to people ...'

There is no room for Johnny Byron in our crowded, buttoned-down land. He answers a need – expressed by the young people who flock to him – but it is a need that society cannot accommodate. The tragedy at the heart of the play is that the world cannot make room for him, just as it can no longer make room for the raids and lion hunts of the *moran*. Much as we might yearn for the life he leads, much as the death of the raw spirit that moves him impoverishes us, he is too big for the constraints within which we have a moral duty to live, the confines which, as Wesley discovers, seem to crush the breath out of us.

There are several ways in which I could try to show that we feel the loss of the wilder life we evolved to lead. I could discuss the urge to shop as an expression of the foraging instinct; football as a sublimated hunt; violent films as a remedy for unexorcized conflict; the pursuit of ever more extreme sports as a response to the absence of dangerous wild animals; the cult of the celebrity chef as an attempt to engage once more with the fruits of the land and sea. The connections in these cases are plausible, unprovable and mundane. I think I have found a more interesting line of evidence.

5

The Never-spotted Leopard

Truly men hate the truth; they'd liefer
Meet a tiger on the road
                    Robinson Jeffers
                    Cassandra

Y iscuid oet mynud
Erbin cath paluc
Pan gogiueirch tud.
Puy guant cath paluc.
Nau uegin kinlluc.
A cuytei in y buyd
Nau ugein kinran
The Black Book of Carmarthen, c. 1250
                    Cath Paluc

The setting was unimprovable. Across the fields, Maiden Castle, a turreted fortress of living rock, clawed at the sky. Beyond it was the village of Wolf's Castle – Casblaidd – distinguished as one of only twenty places in which Owain Glyndŵr was born (he died in quite a few as well), and said to be the spot where the last wolf in Wales was killed. Below us a tangled sallow carr smothered the valley.

'This gap in the hedge here: that could be where it came through. Then it came down the bank, sauntered across the road and disappeared into the scrub.'

I peered into the carr on the other side of the lane. The trees were hooded with ivy. Their mossy trunks sprawled over the ground, or leant on each other, dark-cowled, like drunken friars. Beneath them was an impenetrable thicket of brambles and ferns.

'You wouldn't see him in there, would you?'

'You have no doubt about what it was?'

Michael Disney looked around, at the high bank down which it had come, the narrow strip of pitted tarmac, the low, twisted woodland, and shrugged.

'It's not an issue for me. I saw what I saw and that's that. People can either believe it or not. I'm not trying to convince anyone.'

'You work for the council's public protection division. Has anyone accused you of drumming up business?'

'No, it's not my remit. I'm in trading standards. In fact it's not really anyone's remit.' He smiled slightly, as if picturing the job description. 'What would be the reason for me to put myself in a situation where I could be ridiculed and mocked? I would get nothing from it at all, except a slight bit of notoriety.'

Michael had been driving down the lane towards the A40, returning from an inspection visit. He had heard the stories, seen pictures in the local paper of the prints found at Princes Gate, a few miles to the other side of Haverfordwest, and had not believed a word of it.

'If I'd been dreaming or thinking about them at the time, it might have been another matter. But it was the last thing on my mind. I was just driving along – and one crosses the road. He was probably about three feet high and six feet long. I would say bigger than a medium-sized dog, but definitely not a dog. He was powerful-looking, with a black, glossy, shiny coat, incredibly muscular, like a horse's shoulders. But it was the head that was really strange-looking. I've never seen a head like that, not even in a zoo.'

Michael Disney, former policeman, county council officer, had, to his own astonishment, become one of roughly 2,000 people who see a big cat in the wild in Britain every year.

By the time Michael saw the beast now known as the Pembrokeshire Panther, there had, according to *Wales on Sunday*, been ten 'confirmed sightings'.[1] Some of those who claimed to have seen it were farmers or farmworkers, familiar with the county's less exotic wildlife. Among them were the farmer and – independently – his wife, whose land bordered the lane in which we stood. All described it, as Michael had done, as huge, jet-black, glossy, with a long tail, definitely a cat. One person claimed to have seen it with a lamb in its mouth. Another described how it 'cleared a hedge like a racehorse'.[2] It was blamed for the grisly carcasses of sheep and calves found in remote corners of the farms.

But it was only when the former policeman reported it to both his current and former colleagues that the beast began to be taken seriously. The *County Times* described his sighting as '100% authentic'.[3] Three weeks later, when five people saw it at Rudbaxton, the police sent out an armed response unit. A spokesman for Dyfed-Powys police told me that they were advising people to keep their distance if they saw the Pembrokeshire Panther, and to report it to the council. 'We have to take it seriously, even though strictly speaking it's not a police matter, unless people are in imminent danger.' He added that, in response to reports like Michael's, the Welsh Assembly Government had set up a Big Cat Sightings Unit. I checked: the unit, improbable creature though it is, exists.

I became certain that Michael is an honest, reliable, unexcitable man who has no interest in publicity – in fact he seemed embarrassed by it. I am certain that, in common with other people who claim to have spotted the Beast, he faithfully described what he saw. I am equally certain that the Pembrokeshire Panther does not exist.

There is scarcely a self-respecting borough in Britain which does not now possess – or is not now possessed by – a Beast. Even the London suburbs claim to be infested with big cats: there is a Beast of Barnet, a Beast of Cricklewood, a Crystal Palace Puma and a Sydenham Panther. There have been occasional reports of mysterious British cats throughout history. The earliest written record – *Cath Palug* (Palug's Cat or the Clawing Cat) – is found in the *Black Book of Carmarthen*, written, as the panther runs, thirty miles from where Michael Disney saw his creature. The fragment at the top of this chapter is all that remains of this account: 'His shield was ready/Against Cath Palug/When the people welcomed him./Who pierced the Cath Palug?/Nine score before dawn/Would fall for its food./Nine score chieftains.'[4] But the same animal also appears in the *Welsh Triads*, where its reported attributes present an even stiffer challenge to biology: it was born, alongside a wolf and an eagle, to a giant sow.

Over the past few years the sightings have boomed. In her wonderful book *Mystery Big Cats*, Merrily Harpur finds that 'cat-flaps', as she calls them, are occurring at the rate of 2,000–4,000 a year.[5] As I have discovered while travelling around the country, many others who have not seen these cats ardently believe that they exist.

Among the Beast-spotters are people even better placed to know what they are seeing than Michael and the Pembrokeshire farmers: gamekeepers, park rangers, wildlife experts, a retired zookeeper. As Merrily Harpur notes, around three-quarters of all the cats reported are black, and they are commonly described as glossy and muscular. She also makes the fascinating observation that while the most likely candidate is a melanistic leopard (the leopard is the species in which the black form, though rare, occurs most often), she has not been able to find a single account of an ordinary, spotted leopard seen in the wild in Britain.

Though the sightings are consistent and the witnesses reliable, the hard evidence for an extant population of big cats in the UK is no stronger than the evidence for the Loch Ness monster. In other words, despite the thousands of days cryptozoologists have spent hunting the Beast, despite the concentrated efforts of the police, the Royal Marines and government scientists, there is none.

Though some species of large cat are among the shyest and most cunning of all wild animals, finding evidence that they exist is not difficult, for those who know what they are doing. They are creatures of regular habits. They have territories, dens in which cubs are raised, spraying points and scratching posts. They scatter prints, spraints and hairs wherever they go: the first are immediately recognizable, the provenance of the second and third can be confirmed by DNA testing.

Even those which are seldom seen leave so much evidence that they can be closely studied. I once spent a few days with some biologists in a forest reserve in the Amazon. At night we would hear the jaguars mewing; but I was told by the team leader that, though they might be watching us, we would never see them. One day I wandered down to the stream a few yards from the camp to swim. I spent twenty minutes in the water, then walked back along the sandy path. In my footprints were the pugmarks of a jaguar.

The 2008 Wildlife Photographer of the Year competition was won by a photograph of one of the world's most elusive animals – the snow leopard – taken in one of the world's least accessible places: the Ladakhi Himalayas, 13,000 feet above sea level. The photograph did not just document the existence of the leopard: after thirteen months of experiments, and hundreds of less satisfactory pictures of his quarry, Steve Winter, through a cunning arrangement of camera traps and lights, eventually produced a perfectly composed portrait. 'I knew the animal would come;' he reported. His equipment 'was just waiting for the actor to walk on stage and break the beam'.[6]

Yet, despite camera traps deployed in likely places throughout Britain, despite the best efforts of hundreds of enthusiasts armed with long lenses and thermal-imaging equipment, we have yet to see a single unequivocal image captured in this country. Of the photographs and fragments of footage I have seen – the best the champions of these mysterious felines can produce – around half are evidently domestic cats. Roughly a quarter are cardboard cut-outs, cuddly toys, crude photoshopping or – as the surrounding vegetation reveals – pictures taken in the tropics. The remainder are so distant and indistinct that they could be almost anything: dogs, deer, foxes, bin liners, yetis on all fours. One of the most intriguing features of this story is that hardly anyone who has set out to find a big cat in Britain has ever seen one. Almost without exception, the sightings have been unexpected; in most cases the cats appear to people who had never thought about them or did not believe in them. Pasteur's maxim – that chance favours the prepared mind – seems in this case not to apply.

Nor have the tireless efforts to catch or kill these animals yielded anything more convincing. As Harpur notes, 'more effort and expense than ever went into Imperial tiger hunts has been expended in the hunt for anomalous big cats', and it has produced nothing except a few hapless creatures which have escaped from zoos or circuses or private collections, and are in almost all cases caught within a few hours of their flight. There is a marvellous account in Harpur's book of a policeman sent out at night to investigate the sighting of a lion in Leamington Spa. He stopped to ask a milkman if he had seen the animal. As he did so, he recorded, 'the next thing I was aware of was a passing blur and a sudden weight' in the back of the car. 'In one fluid movement the lion had jumped through the back window on to the passenger seat.' It settled down immediately and the officer, not unconscious of its breath on the back of his neck, drove it to the station.

In 1980, following a series of livestock killings, a female puma was caught in a baited cage trap by a farmer in Easter Ross, in Scotland. At first it appeared to be a wild and ferocious beast, snarling and spitting at its captors. But the effect was spoilt once the puma had settled into Kincraig Wildlife Park: Harpur reports that whenever anyone approached her cage, she would start purring and rubbing against the bars. It seems that she was one of a pair released in the Highlands in 1979 by a man about to be sent to prison. The other was later found dead near Inverness.

Since then, though hundreds of such traps have been set, only one large predator has been caught. A cryptozoologist called Pete Bailey, who had spent fifteen years hunting the Beast of Exmoor, entered one of his traps to change the bait and accidentally tripped the mechanism. He was stuck there for two nights, eating the raw meat he had set for the cat, before he was rescued.[7] We hunt the Beast, but the Beast is us. LOTF

That is about the extent of it: no photos, no captures, no dung, no corpses (except a couple of skulls, which later turned out to have gone feral after they had escaped from a leopardskin rug and a wall trophy), not even a certain footprint. The Beasts of Britain have evaded a five-week hunt by the Royal Marines, police helicopters and armed response teams (it beats logging car crime), a succession of big cat experts and bounty hunters and the mass deployment of the best tracking, attracting and sensing technologies known to humankind. These techniques have worked elsewhere; not here.

In 1995 the government sent two investigators to Bodmin Moor in Cornwall, where the evidence for big cats was said to be strongest. They spent six months in the field, examining carcasses and footprints, exploring the places where the Beast of Bodmin was spotted and photographed. There is something of the nineteenth-century royal commission about this investigation. The report contains photographs of a strapping fellow with a large moustache and a measuring pole, demonstrating the heights of the natural features on which the creatures were photographed.[8] The text reads in places like the final chapters of *The Hound of the Baskervilles*. It is thorough, exhaustive and devastating to those who argued that, while other reputed big cats might not exist, the Beast of Bodmin was real.

They examined the famous video sequence, broadcast widely on television, which shows a cat leaping cleanly over a drystone wall. It looks impressive, until you see the man from the ministry standing beside the wall with his pole, and realize that the barrier is knee-high. A monstrous cat sitting on a gatepost shrinks, when the pole arrives, from a yard at the shoulder to a foot. In one case, where the Beast was filmed crossing a field, and there were no useful landmarks against which to compare it, the investigators brought a black domestic cat to the scene, set it down in the same spot and photographed it from where the video had been taken. The moggie looks slightly bigger than the monster. (Undeterred, the supporters of the Beast of Bodmin now insist that the original pictures show *baby* big cats, whose parents are mysteriously absent from the scene. Stills from these videos continue to be used as evidence that big cats roam Britain.)

The investigators compared a chilling nocturnal close-up of the Beast with a picture of a real black leopard, and spotted an obvious but hitherto unnoticed problem. The panther in the cage, like all big cats, has round pupils, while the creature in the photograph has vertical slits, a feature confined to smaller species, such as the domestic cat.

They examined the three plaster casts of footprints taken from the moor. Two were made by a domestic cat, one by a dog. They attended the gruesome corpses of sheep that local people insisted had been ripped apart by the Beast. That they had been ripped apart was indisputable, but the villains were crows, badgers, foxes or dogs (whose footprints were distributed liberally around some carcasses), and in most cases they had struck after the sheep had died of other causes. While the scientists conceded that it was impossible to prove that a big cat did not exist, they found that there was no hard evidence to support the story. Both the official body Natural England and the Welsh government's Big Cat Sighting Unit, investigating sightings across Britain, confirmed to me that they have come to the same conclusion.

I would go a step further: if a breeding population of these animals existed, hard evidence would be abundant and commonplace. Its absence shows that there is no such population. With the possible exception of the very occasional fugitive (almost all of which have been quickly caught or killed and none of which is black), the beasts reported by so many sober, upright, reputable people are imaginary.

None of this has made the slightest difference, either to the number of sightings or to the breathless credulity with which they are reported in the papers. A story in the *Daily Mail* claimed that 'huge paw prints' in the snow 'could finally be proof' that the Beast of Stroud exists.[9] The woman who found them told the paper 'it looks like someone's just dropped a dart at the end of each toe where its claw has made an indentation in the snow'. This confirms what the photos suggest: the prints were made by a dog. Cats retract their claws when they walk.

A long report in the *Scotsman* titled 'Do giant paw prints mean big cat is on the prowl in Capital?' claimed that marks found by a pensioner in the snow suggest that Edinburgh, like London, is now haunted by a monstrous feline.[10] An 'expert' it consulted decided that 'it's unlikely but not impossible' that the prints were made by a Beast. If so, it must have been a scary creature: a one-legged ghoul hopping up the pavement on tiptoes. Or it might have been someone sticking his fingers in the snow.

There was an equally plausible story in the *Guardian*. It reports the claims of a man who says he was attacked by the Sydenham Panther.[11] The Beast 'jumped on my chest, knocking me to the ground', he said. 'I could see these huge teeth and the whites of its eyes just inches from my face. It was snarling and growling and I really believed it was trying to do some serious damage. I tried to get it off but I couldn't move it, it was heavier than me.' A further report by the BBC alleged that the Panther had him 'in its claws for about 30 seconds', with the result that 'he was scratched all over his body'.[12] Had he really been attacked by a leopard in this fashion, his throat would have been ripped out before he could blink.

My favourite story, from the *Daily Mail*, was headlined 'Is this the Beast of Exmoor? Body of mystery animal washes up on beach'.[13] Beside a photograph of a decomposed head (and another of a snarling black panther), it reported that 'great fangs jutted from its huge jaw, gleaming in the afternoon sun. Then there was the carcass. Up to 5ft long, powerful chest, and what could be the remains of a tail.' The

paper interviewed a local police sergeant, who made the cryptic observation that 'it almost definitely looks like it could be a Beast of
Exmoor'. Only at the bottom of the page did the report reveal that it
was a putrefying seal.

Beast fever has doubtless been heightened by these engaging stories,
but many of those who claim to have seen big cats in Britain also
maintain that they had never heard of them before their own encounter. There is little question that, while a few are hoaxers, most report
their sightings in good faith. In many cases an animal has been seen
by a group of people, all of whom give similar accounts. So what is
going on? Why, over the past three decades, have reports of big cats in
Britain risen from a few dozen a year to thousands?

There is no discussion of this phenomenon in the scientific literature: I cannot find a single journal article on big cat sightings. None
of the psychologists I have contacted has been able to direct me to
anyone studying it.

The fact that most of the reported cats are black perhaps gives us a
clue about what might be happening. Black is the only colour that big
cats of any species commonly share with domestic cats. If you glimpse
what you take to be a ginger leopard or a tortoiseshell lion, you are
likely severely to question your perceptions before allowing yourself
to accept what you think you saw. You are likely to be even more reti-
cent when telling other people about your experience. The mismatch
between colour and size interrupts the process of affirmation, in
which your memory reinforces and perhaps exaggerates what you
saw. The interruption is less likely to occur if the cat is black, which
permits at least the possibility that it could be a panther. The moggie
hypothesis might also explain why no one appears to have seen a
leopard in a leopardskin coat.

Judging the size of an animal is difficult. As David Hambling points
out in the magazine *The Skeptic*, people often imagine that the creatures they see are very much bigger than they are.[14] For example, when
police marksmen cornered an escaped caracal in County Tyrone, they
shot it dead in the belief that it was a lion. Lions are twenty times the
weight of caracals. The Kellas Cat of Scotland is a black beast which
really does exist: it is a hybrid of the Scottish wildcat and the feral
domestic cat. It has often been reported as approximating the size of a

leopard. In fact the biggest specimen ever killed or captured was forty-three inches from nose to tail, which is smaller than the largest wildcats. It may be particularly hard to judge the size of a black animal.

In his book *Paranormality*, the psychologist Professor Richard Wiseman tells us:

> Many people think that human observation and memory work like a video recorder or film camera. Nothing could be further from the truth ... At any one moment, your eyes and brain only have the processing power to look at a very small part of your surroundings ... to help ensure that precious time and energy aren't wasted on trivial details, your brain quickly identifies what it considers to be the most significant aspects of your surroundings, and focuses almost all of its attention on these elements.[15]

The brain, he says, scans the scene like a torch searching a darkened room. It fills in the gaps, to construct what appears to be a complete image from partial information.

This image can then become lodged in our memories, and we treat it as if it were as concrete and definitive as a photograph in an album. If we are focused on a cat and not on its surroundings, it could be that the process of singling out the beast magnifies it and shrinks the setting.

I wonder, too, whether there might be a kind of template in our minds in the form of a big cat. As these were once our ancestors' foremost predators,* we have a powerful evolutionary interest in recognizing them before the conscious mind can process and interpret the image. It could be possible that anything which vaguely fits the template triggers the big cat alarm: we lose little by seeing cats which do not exist, but lose a lot by failing to see those which do.

But none of this explains why big cat sightings appear to have become much more common in recent years. The phenomenon is not

---

* Finding myself in South Africa soon after reading Bruce Chatwin's famous account, I asked a curator at the Transvaal Museum to show me the skulls of *Dinofelis*, the false sabretooth cat, and those of the hominids on which it is believed to have preyed, punctured, just above the spinal column, by its massive canines. They were just as Chatwin described them in *The Songlines*.

confined to Britain, though it appears to be particularly widespread here; there have also been plenty of unlikely sightings in other parts of Europe, in Australia and in areas of North America that long ago lost their cougars and jaguars. Feral domestic cats have lived in the British countryside for centuries, and there is no reason to suppose, and no evidence that I have seen, that a higher proportion of them are now black. It could be, with the decline of gamekeeping, that their population has risen, but that must be offset against the fact that we spend less time outdoors; it seems unlikely that this outbreak of catatonia can be explained by a rising number of encounters with moggies.

Certain paranormal phenomena afflict every society, and these phenomena appear to reflect our desires; desires of which we may not be fully conscious. In Victorian Britain, large numbers of people believed that the dead were appearing to them and communicating with them. They saw ghosts, heard voices and imagined they could exchange messages with the departed through séances and table-turning. The Victorians were obsessed by death. Walk around any ancient graveyard and you will read the tragic story of that era: children and spouses snatched away, sometimes, in the epidemics that raged through the crowded cities, within days of each other. Ours was a nation in perpetual mourning. The notion that the dead could return in this life must have been almost as comforting as the belief that we would be reunited with them in the afterlife. Today reports of contact with the dead are less prevalent.

As the space race between the United States and the Soviet Union gripped the world's imagination, sightings of UFOs and aliens, almost unknown in previous eras, multiplied. This was a period in which we entertained great hopes for the transformative potential of technology, in which large numbers of people fantasized about living on other planets and travelling across galaxies and through time. It was also an epoch in which the world was shrinking, and we were becoming aware that the age of terrestrial exploration and encounters with peoples unknown to us was ending; that planet earth was perhaps a less exciting and more certain place than it had been hitherto. Aliens and their craft filled a gap, tantalizing us with the possibility that encounters with unknown cultures could continue, while promising that we too would achieve the mastery of technology and physics we ascribed to extraterrestrials. Today, perhaps because our belief in technological deliverance has declined, we hear less about UFOs.

Could it be that illusory big cats also answer an unmet need? As our lives have become tamer and more predictable, as the abundance and diversity of nature have declined, as our physical challenges have diminished to the point at which the greatest trial of strength and ingenuity we face is opening a badly designed packet of nuts, could these imaginary creatures have brought us something we miss?

Perhaps the beasts many people now believe are lurking in the dark corners of the land inject into our lives a thrill that can otherwise be delivered only by artificial means. Perhaps they reawaken old genetic memories of conflict and survival, memories which must incorporate encounters – possibly the most challenging encounters our ancestors faced – with large predatory cats. They hint at an unexpressed wish for lives wilder and fiercer than those we now lead. Our desires stare back at us, yellow-eyed and snarling, from the thickets of the mind.

I suppose and I generalize, of course, but the reification of our inner big cats is not the only phenomenon which hints at such yearnings. Consider the widespread and otherwise inexplicable response to the death of Raoul Moat. In 2010, Moat was discharged from Durham Prison after serving a sentence for beating up a child. Armed with a sawn-off shotgun and prompted perhaps by 'roid rage' – the explosive, irrational anger experienced by body builders who take steroids – he set out to settle imagined scores with his former girl-friend and the police. He shot his ex-partner in the stomach and killed her boyfriend, then blinded a policeman by blasting him in the face.

Officers from eight police forces mobilized to capture him, but he evaded them for almost a week, living rough, sleeping in drains and abandoned buildings. At the height of the search, 10 per cent of all the available duty officers in England and Wales were deployed to hunt him. Parts of Northumberland were evacuated. When at last he was cornered, the stand-off lasted for six hours, before Moat shot himself in the head.

He was, in other words, an unlikely hero: child-beater, murderer, mutilator of unarmed people. Yet, long after his death, paeans to

Moat are still appearing on Facebook pages.* Here is a small sample.

> R.I.P Sir Raoul Thomas Moat – A True Peoples Champion. Sir Raoul was murdered in cold blood by Northumbria police, anyone that knows the sound of a shotgun blast will know he didn't kill himself. We will fight to get justice for you our brave fallen soldier.

> R.I.P Raoul You Were A Propa LEGEND ! Ganna Be Missed Mate ! Wish People Were Like You When Said Your Going To Do Something You Mean ! STILL THINK YOU COULD HAVE WENT LONGER ! R. I.P MATE ROCK HEAVEN LIKE YOU ROCK DOWN HERE ! YOU TOTAL LEGEND

> A True Peoples Champion . . . It is sick the way our national treasure has been treated. R. I. P Sir Raoul Thomas Moat, gone but never forgotten.

There are thousands of messages like these, posted by both sexes. Moat seems to have become a vehicle for urges to which we cannot afford to succumb. He is admired for his ability to evade capture, flitting like a wild beast through the brakes and coverts of Northumberland, outfoxing the hounds and helicopters deployed by the police. He had burst from his enclosure and gone feral, and in doing so he appears to have unleashed the desires of people who feel trapped in their lives. Several of the commentators lamenting this adulation for a killer used the same term. They complained that Moat had been 'lionized'.[16] This word carries more weight than the authors intended.

---

* Moat's story – and the strange public response – recapitulates that of Harry Roberts, the armed robber and sadistic murderer of prisoners during late colonial wars, who went on the run in 1966 after shooting dead two policemen. He hid in the woods for ninety-six days before he was captured. Like Moat, this revolting man was celebrated by some people as a folk hero.

6

Greening the Desert

When through the old oak forest I am gone,
Let me not wander in a barren dream
John Keats
On Sitting Down to Read King Lear Once Again

All Hallows' Eve. *Nos Galan Gaeaf*. Early frosts and still days had engineered a blazing autumn. The birches looked like a shower of gold coins. An occasional beech tree flamed against the pale ash leaves and the mauve-brown oaks. The sun was a pewter gleam behind the clouds, the air was almost still. There was a thickness to the day, as if it had been laid on with oil paint, or as if air and leaf and ground were the flesh of a single organism. The berries of the hawthorn exuded from the woods like specks of blood.

Beside the track the dying willowherb had sprung white whiskers. Rills trickled through saxifrage and honeysuckle. Late caddis flies rose from the water and oared the thick air. From across the valley I heard an ancient sound, now rare in these hills: a farmer calling and whistling to his dogs. I left the path and stepped up into the last scrap of woodland before the desert began.

The woods climbed a gentle slope. As I walked towards the light, sheep clattered away from me. I startled a jay and a great spotted woodpecker, which swooped off through the autumn trees with a long, high note. The forest floor had been scrubbed clean. Beneath the fallen leaves there was nothing but moss, sheep shit and mud. A single wood hedgehog mushroom had been turned over by the sheep, and showed its long fine teeth. There were no leafy plants, no saplings, no tree younger than around a century, no understorey of any kind. Many of the oaks had fallen or were close to death. The old wood was dying on its feet. By eating all the seedlings that raised their heads, the sheep were killing it.

The wood petered out into birches, bracken and the odd rowan tree, then into spongy pastures. As I walked up the bare hillside, I could see the mossy domes where trees had fallen: the burial mounds of what had until recently been a larger forest. I hacked through bracken and yellow grass and over anthills covered in red moss. The bracken soon gave way to moorgrass, now greying after the sharp frosts. The last of the waxcap and *Inocybe* mushrooms had flopped over on their stems.

I climbed to the top of a small hill. To my east was Bryn Brith, the speckled hill, whose name suggests that it lost its trees long ago. The yellow grass was still mottled with patches of blue-green gorse. Beyond it were the long blurred slopes of the hills surrounding Pumlumon, the highest mountain on the plateau, grey-brown and treeless. To the south, the hills graded from yellow to green to blue as they stepped away, deep into Ceredigion and Pembrokeshire. Beyond them I could glimpse a grey blur of sea.

Though I could see for many miles, apart from distant plantations of Sitka spruce and an occasional scrubby hawthorn or oak clinging to a steep valley, across that whole, huge view, there were no trees. The land had been flayed. The fur had been peeled off, and every contoured muscle and nub of bone was exposed. Some people claim to love this landscape. I find it dismal, dismaying. I spun round, trying to find a place that would draw me, feeling as a cat would feel here, exposed, sat upon by wind and sky, craving a sheltered spot. I began to walk towards the only features on the map that might punctuate the scene: a cluster of reservoirs and plantations.

Out of the woods, the day felt colder. It had seemed still among the trees. Here there was a cutting, damp wind. I followed a path that took me along the line of a fallen drystone wall, now replaced with posts and wire. No bird started up – not even a crow or a pipit. There were neither fieldfares nor redwings, larks nor lapwings. With the exception of the chemical monocultures of East Anglia, I have never seen a British landscape as devoid of life as the plateau some local people call the Cambrian Desert. In most places the nibbled sward over which I walked contained just two species of flowering plant, the two that sheep prefer not to eat: purple moorgrass and a small plant with jagged leaves and yellow flowers called tormentil.

I followed the Bwlch-y-maen – rocky hollow – trail over bare hills and down bare valleys until it brought me to a point overlooking a wide basin, cradling a small reservoir called Llyn Craig-y-pistyll. I sat on a rock and felt myself slumping into depression. The grass of the basin was already dressed in its winter colours. There were no tints but grey, brown and black: grey water, cardboard-coloured grass, a black crown of Sitka spruce on the far hills. The occasional black scar of a farm track relieved rather than spoilt the view. My map told me that if I walked for the rest of that day and all the next, nothing would change: the plateau remained treeless but for an occasional cluster of sallow or birch, and the grim palisades of planted spruce.

As I glared at the view, the weather front passed in a litter of cloud-lets and the sun broke through. Far from enlivening the scene, it brought the bleakness into sharper focus. Now I could see the grey wall of the spruce trunks and the green battlements that surmounted them. The emptiness appeared to expand in the sunlight. I trudged down to the lake. Five Canada geese sat on the far bank, the first birds I had seen since leaving the woods, two hours earlier. They waddled into the water when they saw me, and floated away, grunting softly. Sheep scoured the far bank.

The water was surprisingly low for autumn, exposing the shaley rubble of the banks and the black mud of the reservoir floor, rutted with sheep tracks. I sat by the water and ate my lunch. From where I sat, the tops of the spruce trees looked like an approaching army edg-ing over the hill, pikes raised. I realized that, though this was a Sunday, I had not seen a soul. I leant against the exposed bank of the reservoir, mentally dressing the land, picturing what might once have lived there, what could live there again. Then I rose, stumbled up the hill and ran back along the track. When I returned to the glowing hearth of the ruined wood, with its occasional bird calls, I almost wept with relief.

The Cambrian Mountains cover some 460 square miles, from Machynlleth in the north to Llandovery in the south, Tregaron in the west to Rhayader in the east. They are almost uninhabited, almost unvisited: two friends of mine once walked across them for six days without seeing another person. They begin 300 yards from my home. I see them from my kitchen window, rising through *fridd** and birch woods to a bare skyline.

Before I moved to Wales, I lived for several years in a densely peo-pled quarter of a city. Whenever I heard the wild cry of gulls, and looked up to see them crossing the narrow strip of sky, I felt a small tear in the cloth of my life elongate a little more. At those moments I knew that I was in the wrong place. Where they were going, I wanted to be.†

When I arrived in Wales, and found myself living between two of the least-inhabited places in Britain – the Cambrians on one side of my valley, Snowdonia on the other – I felt almost overwhelmed by choice. Like a battery chicken released from its cage, at first I ventured into the mountains tentatively, not quite believing that I could step out of my front door and walk where I would for as far as I wanted, and seldom encounter a road or a house.

But as I began to explore these great expanses, often walking all day over the hills, my wonder and excitement soon gave way to disappointment; the disappointment gave way to despair. The near-absence of human life, I found, was matched by a near-absence of wildlife. The fragmented ecosystems in the city from which I had come were richer in life, richer in structure, richer in interest. In mid-Wales, I found, the woods were scarce and, in most cases, dying, as they possessed no understorey. The range of flowering plants on the open land was pitiful. Birds of any kind were rare, often only crows. Insects were scarcely to be seen. I have walked these mountains for five years now, and with the exception of a few small corners, found no point of engagement with them. Whenever I venture into the Cam-brian Desert I almost lose the will to live. It looks like a land in perpetual winter.

---

* *Fridd* is the land between the enclosed fields of the valley bottoms and the open moor at the top of the hills. It tends to cover the steep slopes of the hillsides and to be dominated by scrub and bracken.
† Unless it was the municipal rubbish dump!

It is seen as disloyal, especially in this patriotic nation, to talk the landscape down. Some people say they find it beautiful. The Cambrian Mountains Society celebrates its emptiness. It describes the region as a 'largely unspoiled landscape',[1] and approvingly quotes the author Graham Uney, who claims, 'there is nothing in Wales to compare to the wilderness and sense of utter solitude that surrounds these vast empty moorlands'.[2] To which I say, thank God. What he extols as wild, I see as bleak and broken. To me these treeless, mown mountains look like the set of a post-apocalyptic film. Their paucity of birds and other wildlife creates the impression that the land has been poisoned. Their emptiness appals me. But I also recognize that it is a remarkable achievement.

For the Cambrian Mountains were once densely forested. The story of what happened to them and – at differing rates – to the uplands of much of Europe is told by a fine-grained pollen core taken from another range of Welsh hills, the Clwydians, some forty miles to the north.[3] A pollen core is a tube of soil extracted from a place where sediments have been laid down steadily for a long period, ideally a lake or a bog in which layers of peat have accumulated. Each layer traps the pollen that rains unseen onto the earth, as well as the carbon particles which allow archaeologists to date it.

The Clwydian core was taken in 2007 from a mire in which peat has settled for the past 8,000 years. At the beginning of the sequence, the plant life was still affected by the cold, dry conditions following the retreat of the ice. Trees – hazel, oak, alder, willow, pine and birch – accounted for about 30 per cent of the pollen in that layer, grass for much of the rest. As the weather became wetter and warmer, elm, lime and ash trees started to move in. The woods became deeper and darker. By 4,500 years ago, trees produced over 70 per cent of the pollen in the sample. Heather pollen, by contrast, supplied around 5 per cent.[4]

Farmers began to colonize the hills in the Neolithic period (between 6,000 and 4,000 years ago). Over the millennia, they gradually cleared some of the land for crops, ran their sheep and cattle on the hills and burnt the remaining trees. The clearing and burning and grazing stripped the fertility of the soil, encouraging heather – which thrives on poor land – to grow. Until some 1,300 years ago the peat still contained pollen from most of the trees of the ancient wildwood. The ash and elm disappeared from the sequence soon afterwards, then the lime and pine, then – but for a few relict stands – the other species.

As the trees retreated, the heather pollen began to rise. The pollen core marks a brief recovery of forest during the plague and economic collapse of the fourteenth century, and the turmoil caused by Glyndŵr's revolt in the fifteenth century. But the regeneration did not last long. By 1900 the proportions of 1,000 years before had been inverted: trees supplied just 10 per cent of the pollen in the core, heather 60 per cent. The forest had been replaced by heath. Over much of the British uplands today, particularly the Cambrian Mountains, the heath has now given way to grass.

Heather took longer to dominate the Clwydian Hills, where the soil is relatively fertile, than most of the uplands of Britain. Where the soil was thinner, it became the dominant vegetation as early as the Bronze Age, between 4,000 and 2,700 years ago. I think of the Bronze Age as the period in which the hills turned bronze.

This record, and similar evidence from the rest of the country, shows us several things. It shows that the open landscapes of upland Britain, the heaths and moors and blanket bogs, the rough grassland and bare rock which many people see as the natural state of the hills, which feature in a thousand romantic films and a thousand advertisements for clothes and cars and mineral water, are the result of human activity, mostly the grazing of sheep and cattle. It shows that grazing and cultivation have depleted the soil. It shows that when grazing pressure eases, trees can return.

The word woodland creates a misleading impression of what the ecosystem of these hills would have looked like after trees returned in the early Mesolithic, and until they were cleared by farmers. From Scotland to Spain, the western seaboard of Europe was covered by rainforest. Rainforests are not confined to the tropics. They are places wet enough for the trees to carry epiphytes, plants which grow on other plants. A few miles from where I live I have found what appears to be a tiny remnant of the great Atlantic rainforest, a pocket of canopied jungle, protected from sheep, in the Nantgobaith gorge. The trees hanging above the water are festooned with moss and lichen. Polypody – the many-footed fern – slinks along their branches.

Through the forest canopy move troops of long-tailed tits, goldcrests, nuthatches and treecreepers. Walking up Cwm Nantgobaith one autumn day, I noticed something unmistakable, but so unfamiliar that it took me a moment to process it. It shone like a gold sovereign against the brown oak leaves on the path. I picked it up.

It was a leaf of *Tilia cordata*, the small-leaved lime. Daffodil yellow, onion-shaped, it filled only the indentation in my palm. I looked up the path and saw another, then another. I followed the trail to two great trunks, forking from one stool and twisting up into the canopy above the path. I had walked beneath them many times but never noticed them: swaddled in deep moss, the trunks were indistinguishable from those of the oaks, and the leaves appeared only far above my head. Since then I have found several more limes in the gorge. This is a tree of the ancient wildwood which is now rare in Wales. Its presence there suggests that this fragment of rainforest might have grown without interruption since prehistoric times.

Heather, which many nature-lovers in Britain cherish, is typical of the hardy, shrubby plants which colonize deforested land. I have seen similar landscapes of low scrub in Brazil, Indonesia and Africa, where logging, burning and shifting cultivation have depleted the soil. I do not see heather moor as an indicator of the health of the upland environment, as many do, but as a product of ecological destruction. The rough grasslands which replace it when grazing pressure further intensifies, and which are also treasured by some naturalists, are strikingly similar to those whose presence we lament where cattle ranching has replaced rainforests in the tropics. I find these double standards hard to explain. I wonder whether our campaigns against deforestation elsewhere in the world, commendable as they may be, are a way of not seeing what has happened in our own country.

This is not to say that there was no open land. In some places the soil was too poor or wet for trees to grow. On the tops of the highest mountains the weather was too cold and harsh. But these open habitats were small and occasional, by comparison to the great tracts of wildwood which covered most of the hills.[5] Nor is this to suggest that if human beings and their domestic animals were suddenly to vanish from Britain, our ecosystems would soon revert to those that prevailed in the Mesolithic. The uplands have been so depleted of nutrients* and their soils so compacted by sheep that they are unlikely to support continuous forest. For a few centuries after rewilding began, they would be more likely to host a patchwork of rainforest, covert, scrub, heath and sward.

[The ancient character of the land, the forests that covered it and the animals that lived in them – which until historical times included wolves, bears, lynx, wildcats, boar and beavers – have been forgotten by almost everyone.]The open, treeless hills are widely seen as natural. The chairman of a trade association called Cambria Active describes the scoured acid grassland it is trying to promote to tourists as 'one of the largest wildernesses left in the UK'.[6] The Countryside Council for Wales, the nation's official conservation agency, calls its Claerwen nature reserve, a bare waste of sheep-scraped misery in the Cambrian Mountains, 'perhaps the largest area of "wilderness" in Wales today'.[7]

Spend two hours sitting in a bushy suburban garden anywhere in Britain, and you are likely to see more birds, and of a wider range of species, than you would while walking five miles across almost any open landscape in the uplands. But to explain that what we have come to accept as natural is in fact the aftermath of an ecological disaster – the wasteland which has replaced a rainforest – is to demand an imaginative journey that we are not yet prepared to make. Our memories have been wiped as clean as the land.

There is a name, coined by the fisheries scientist Daniel Pauly, for this forgetting: 'Shifting Baseline Syndrome'.[8] The people of every generation perceive the state of the ecosystems they encountered in their childhood as normal. When fish or other animals or plants are depleted, campaigners and scientists might call for them to be restored to the numbers that existed in their youth: their own ecological baseline. But they often appear to be unaware that what they considered normal when they were children was in fact a state of extreme depletion. In the uplands of Britain, naturalists and conservationists bemoan the conversion of heather into rough grassland, or of rough grassland into fertilized pasture, and call for the ecosystems they remember to be restored – but only to the state they knew.

---

* Nutrients are lost as animals are removed from the land for consumption in other places, and as soil is leached or stripped by erosion.

The main agent of these transformations is an animal which, like the flayed hills, we have come to accept as part of the fabric of British life: a woolly ruminant from Mesopotamia. No wild animal resembling the sheep has ever existed in Britain or western Europe. (The musk ox, which belongs to the same sub-family as sheep and goats, probably comes closest, but it has a different ecology and set of habitat preferences.) The mouflon, the 'wild' sheep of Corsica and Cyprus, is in fact one of the earliest examples of a feral invasive species: a descendant of animals which escaped from domestic herds during the Neolithic.[9]

Because they were never part of our native ecosystem, the vegetation of this country has evolved no defences against sheep. In the uplands they rapidly deplete nutritious and palatable plants, leaving behind a remarkably impoverished flora: little beside moss, moorgrass and tormentil in many places. The sheep has caused more extensive environmental damage in this country than all the building that has ever taken place here.

The horses watched as wild animals watch, ears pricked and turned towards us, eyes locked, occasionally tossing their heads and snorting, ready to flee. But when we squatted down and waited, they began to move towards us. A careless movement scattered them. They swirled away then stopped a little distance off, regrouped, edged towards us again, chewing, snorting, stamping, tossing. So powerful did their curiosity seem, so much more powerful than their evident fear, that it was almost as if, like us, they craved this contact with another species.

A wisp of wind blew over us towards them. They twitched and flared their nostrils, tubes of muscle flexing all the way up their long faces. I was struck by this thought: that if you landed on an unknown continent and saw the mammals or birds that lived there, you could tell immediately whether they were predators or prey. The eyes of the eaten are on the sides of their heads, as they need a wide range of vision. The eyes of the eaters are at the front, as they need to focus to catch their quarry.

Ritchie had brought me to the land on which he had once lived, and where, with others, he had shut out the sheep and begun replanting, twenty years before. It was a cool, still morning, the first day of autumn, a year after my dispiriting foray into the Desert. On the other side of the fence the birch and ash still held their paling leaves. The hawthorns and rowans were already bare. Far below us, in the remnant stand of mossy oaks that grew beside the stream in the sheep pastures, jays screeched like football rattles.

Ritchie Tassell is the person to whom I have most often turned when trying to feel my way through this story. He has a voracious appetite for reading, and made some of the key discoveries in the literature that feature in this book. More importantly, he has an engagement with the natural world so intense that at times it seems almost supernatural. Walking through a wood he will suddenly stop and whisper 'sparrowhawk'. You look for the bird in vain. He tells you to wait. A couple of minutes later a sparrowhawk flies across the path. He had not seen the bird, nor had he heard it; but he had heard what the other birds were saying: they have different alarm calls for different kinds of threat.

He was brought up in a village in Northamptonshire – its burr still lingers – the county,whose wildlife and human life were celebrated by the poet John Clare, who died a century before Ritchie was born. His grandfather often took him out into the fields and woods, teaching him about birds. 'He showed me how to summon owls out of the trees. It's been a party trick of mine since I was about eight.'

His grandfather studied at Kettering grammar school at the same time as the author H. E. Bates; they both came from humble shoe-making families.

'My grandad and my father avidly read his books, which often recalled his childhood in the Northamptonshire countryside. Listening to them talk, I began to realize the great losses my grandad's generation had witnessed in their own lifetime.'

Ritchie is obsessed with birds and for that reason, he says, he can seldom watch a television drama. 'There's this hideous habit in which British films are overdubbed with American bird tracks. They're obsessive about the setting, the period costumes, the hair, the vehicles, the horses, but they always get the birdsong wrong. I've got to the point where I have to leave the room. I cannot stand it: it's a measure of how disengaged we are. We could probably as a nation lose all our birds and there's an increasing number of people who wouldn't even notice.

'As we become more urban we're losing our attachment. Many of our summer migrants could just slip away and most of us wouldn't know it. To me that's shocking.'

When Ritchie was a small child, Dutch elm disease reached the land around his village. 'We had 300-year-old sentinel elms which dominated the landscape for miles. I remember the gangs of timber cutters turning up and felling the trees and burning the roots. What I considered to be the permanence of the countryside suddenly wasn't.

'I managed to persuade myself that it was a natural tragedy. But soon afterwards, in the 1970s and early 80s, something even worse happened. The mixed farms started going down the pan, and agribusiness began to take over. The farmer next door was one of the last to go, he still had cattle and sheep and arable crops in rotation. A week after he sold up to a big pension fund this fleet of bulldozers arrived. They completed the job that Dutch elm disease had begun. They stripped the hedgerows, the remaining parkland trees, walnut trees two or three hundred years old: the whole lot was gone in a day.

'That's where I got my environmental consciousness from. I was about twelve at the time. Seeing how it can all disappear at our whim, the shock of seeing this entire landscape being erased. The old farmer probably had half a dozen full-time staff. You could see them every morning walking across the fields. It all went almost overnight. From then on everything was done in fleets of big tractors. As the combines left the field, the subsoilers would move in, then the ploughs. It was like a military operation.

'That was the worst of times in terms of habitat destruction, almost the final nail in the coffin of what John Clare was writing about. He was there at the beginning of the process, I was there at the end. It was a permanent loss. It's all gone.'

As part of his first degree, Ritchie took a placement at the Centre for Alternative Technology in Machynlleth.

'It all came together in my head: the care for the land and our impact on it, the importance of minimizing it. After working in London, I moved back to Wales in the early 90s and got a job as a carpenter on the cliff railway at the Centre. I started working as a contractor managing small-scale woods. I fairly quickly realized that if I was going to pursue that I'd have to go back to college – I took a masters in environmental forestry. After that I got a job as a woodland officer and I've been doing it ever since. It didn't take me long to see that the most radical thing you could do round here was to put fences around the woods and keep the bloody sheep out.'

The land to which he had brought me belonged to a communal house in which he had once lived. It had its own hydroelectricity supply and a plan, which had been hatched before he moved in, to buy some of the surrounding land and plant trees.

The Cambrian Mountains must be among the most unpromising places in northern Europe for a rewilding experiment. Grazed and cleared for thousands of years, infertile, naturally acid and further acidified by pollution from power stations, scourged by wild Atlantic storms and almost constant wind, they look as if they could sustain no more than the mangy pelt with which they are now clothed. But, starting with a treeless sheep pasture high above the estuary, Ritchie had begun to discover what worked and what failed.

As we moved through the young woods, a troop of blue tits, coal tits and long-tailed tits followed us, working through the branches, grating and cheeping, picking tiny insects from the cracks in the bark. The trees, Ritchie told me, had not taken easily. When the sheep were shut out, bracken and coarse grasses had sprung up, through which the seedlings had struggled to establish themselves. To accelerate the process, in some places he and his friends had turned the turf over with mattocks. In others he had cut the bracken every summer, so that it would not flop over the seedlings as it died, smothering them.[10] Now the trunks of the trees were as thick as my calf, and they towered over us: the tallest were perhaps twenty feet high.

'Somehow,' Ritchie told me, 'I didn't think I'd live long enough to see this.'

Though he is a little younger than me, I understood that. Walking in the Cambrian Desert, it sometimes seems impossible to imagine trees returning there: the emptiness stands as an incontestable fact, as if it were a matter of geology, not ecology. Yet here, where local farmers had told him that trees would never grow, this sedimentary law had been reversed. The habitat through which we ducked now qualified as a wood, and it was already hard to picture the sheep pasture that had preceded it.

They had planted trees, but soon discovered that, in much of the fenced land, this was unnecessary. Where they had turned over the turf, the exposed soil was colonized by birch seed, which blew in from a few surviving trees further down the valley, which had themselves returned, Ritchie explained, as a result of an agricultural depression around a century ago.

'Almost every tree we planted has now been overwhelmed by native birch. It grew so densely it looked like the cress you grow on your windowsill. Even when the trees we planted survived, the local birches did much better. They're genetically suited to this site. Seeing the way the birch recolonized was a real awakening. I saw that nature is far more adept at doing these things than we are.'

Ritchie's experiments, which became the basis of his master's dissertation, demonstrated that birch could be sown, with the help of some scraping of the soil and hacking of the ferns, into dense bracken, without the need for herbicides.

'It's all about soil disturbance with birch. It's designed to chase retreating glaciers and ice sheets by seeding into the exposed soils before the coarse grass gets a foot in the door. It's also good at recolonizing burnt sites and places where the conifers have been felled. You just need to prepare the site with a tractor or a rotovator. Or you could use cattle or pigs or wild boar to break up the bracken and disturb the soil. If we're serious about getting forests back in the uplands as quickly as possible, this has to be a way to go. It could work out a lot cheaper than planting and weeding nursery stock.'

In the acid hills, he told me, birch, with its slightly alkaline leaf litter, prepares the soil for other trees. At the foot of the twisted black and white trunks, orange toadstools – birch boletus – grew. They looked like soggy bread rolls, or, in the green rockpool light beneath the trees, like sea sponges. They pushed their way through dead leaves, deep moss, bilberry and little ferns. In some places the big soft leaves of foxgloves flopped over the ground. It was hard to grasp that this land had belonged to the Desert just twenty years before.

We scrambled across the slope until we reached a treacherous band of exposed rock, over which grew algae and slimemold, like the first land plants to colonize the Earth. Ritchie told me that there had been a landslip here ten years before: the thin soil had slid off a sheet of polished rock. Now alder, sallow and birch had colonized the exposed earth, and their roots were fingering across the stone, gathering soil, stabilizing the slope. He had planted an aspen on the edge of the slip: it was suckering up around the rock. Its leaves, the shape of the domes on a Russian Orthodox church, never quite still, shivered in the cool bright light. Brown leathery flanges of cup fungus grew in the flushes developing around the exposed stratum. Jays screeched among the trees. They will become, if the land is defended from sheep for long enough, one of the agents of reforestation. Jays can each bury 4,000 acorns every autumn, sometimes miles from where the mother tree grows. While, astonishingly, they can remember where they put every one, some of the birds will die in the winter, allowing the seeds to germinate.[11]

We crossed the fence again, and stepped up into the adjoining pastures. Ritchie explained that the farmer had stopped running sheep here, instead keeping horses on the land, which appeared to have been left to go wild. He had found the skeleton of a foal in the grass: the animals seemed to be looking after themselves. A few small rowans had slowly begun to establish themselves on the hillside. Their silvery trunks caught the light. We stood above the young wood in its early autumn colours, looking down the valley whose bluffs interlocked like the fingers of two hands, falling away to the estuary, beyond which Snowdonia rose into that crystal day.

From behind us, like a dark bolt fired through the back of our minds, a peregrine appeared, high against the wisps of cirrus. It swept across the sky without moving its wings, in one smooth, swift glide which seemed to follow the curve of the earth. It turned above the far hill, whereupon a kestrel appeared, sliding down a column of air to attack. With a flick of its wing, the larger falcon swung away, soon diminishing to a speck high above the estuary.

The troop of tits caught up with us, moving through the tops of the trees below our heads. They filled the wood with their noise, squeaking and churring like an unoiled wheelbarrow. Where the horses had skidded on the wet grass, scarring the pasture, hedge bedstraw and wood sorrel grew, relics of some ancient woodland edge. We could see the animals on the other side of the little valley, the foals grazing at the heels of the mares.

The hills on the far side of the estuary were now patched with the small dark shadows of cloudlets, the scouts deployed by the great battalions massed at the offing as a front approached across the sea. A young buzzard soared above the horses then began to mob the kestrel.

We walked through the pastures around the top of the wood, stumbling across a little waterfall, sudden and surprising in the midst of bracken and gorse. Marsh marigold leaves withered on the banks.

'This,' Ritchie said, 'is the end of life as we know it. From here on up there are no trees, except for that one birch.'

I looked up for a moment at the bare, bleak plateau, the pony paths converging into the distance, the hessian emptiness, then turned away.

We climbed back over the fence and stood among the trees he had planted at the topmost corner of his old land. Here the soil was thin and poor. He had found little piles of stones – about the size of fists – gathered together, which suggest that it had been cultivated. Ritchie told me that he had once met an old man in the local market who, in the 1940s, was part of a team of contract mowers working with scythes, travelling from farm to farm during the harvest. They had come to this farm to harvest the oats, in fields further down the valley. 'It was a privilege to meet him. He was the last of his kind, and the harvest here was one of the last he ever cut.' But this land, high in the watershed, might not have been tilled for many hundreds of years: the piles of stones, Ritchie said, could date back to the Bronze Age, when shifting cultivation was practised. 'It was probably similar to slash and burn farming in the tropics. It would have exhausted the soil pretty quickly and they would have moved on.' (The difference, in the tropics, is that the soil and vegetation often recover quickly; the impact of traditional shifting cultivation can be low. In the Cambrian Mountains, probably because nutrients are quickly stripped from the exposed soils by rain, this does not seem to happen.)

The rowans, on this poor soil, had, in twenty years, grown to only four feet. They were wizened and wind-bitten. The oaks had scarcely grown at all; they had put out a few weak branches just above the soil, which were now dying back. But the pines he had planted were twelve feet high. These are Scots pine but, as Ritchie points out, that is a misnomer: they are widely distributed in Europe, and were once widespread in Britain. As the pollen core suggests, pine seems to be welladapted to the tops of the Welsh hills. Forestry and conservation bodies sometimes claim that, outside Scotland, it does not belong to our native flora. But many of the biggest trees in the Bronze and Iron Age fossil forest exposed on Borth beach, close to Aberystwyth, are pines. Beautifully preserved in an ancient peat bog, they still possess their scaly orange bark. A few bilberries clung to their bushes. I tried some: to my surprise they still tasted good. They must have been hanging there for three months.

'Though this might have been deep forest once, with the depletion of the soil on the worst upland sites that's not what all of it would revert to immediately today. We'll get an ecosystem that has never been here before: a mosaic of habitats of different structures and sizes, intricate and diverse. The trees that grow slowly will be grazed more, as they can't get beyond the reach of the animals. It'll probably take many years for the leaf litter to make a reasonable soil here again, allowing other species to move in.'

Two late swallows flickered past, dipping and flexing over the meadows. In the new woods below us I could hear the chatter of siskins. The front now loomed over the valley, casting its shadow on the hills across the estuary, driving the winds before it. I looked over the tops of the trees beneath us and thought what a wonderful thing it must be to leave such a legacy, that the woods Ritchie had planted would stand long after anything I had made or written had vanished from the Earth.

Even so, something was missing.

The day had begun so dark and grim that it seemed as if the sun had taken one look, turned over and gone back to sleep. Now, as it struggled to throw off the ragged counterpane of cloud above the mountains, the raw November day began to brighten up.

The trees had shaken off most of their leaves. A few scraps of russet still clung to the oaks and beeches, but the birch and sallow reclaiming the ground around the pond were now grey smudges against the dead grass. We stood in the mud churned by the raising of the fence – triply secured to reassure local people – waiting. The film

77

cameramen adjusted their tripods and stamped their feet to keep warm. An ecologist uncapped his binoculars. The volunteers – baggy jumpers, torn trousers, dreadlocks and nose rings – smoked roll-ups and spoke in tense whispers. From the larch plantation on the mountain to the west I could hear the distant baying of hounds and the occasional warble of a hunting horn. The still, cold air trickled down my neck.

A monstrous bull mastiff, all saggy skin and jowls, that had been snuffling round our feet, suddenly leapt into the air, squealed like a piglet, then ran whimpering to its owner: it had touched the electric fence.

'I think we're ready to go,' someone said.

Two young men with blond beards wedged hoardings into the mud on either side of a great box. One of them drew out the pins which secured the panel facing the pond. A moment later there was a flash of chocolate fur between the boards, then another: two large animals blurred past and disappeared into a rough hut of sticks and rushes that had been built at the water's edge.

After a few minutes, just as one of the bearded men had promised, the willow branches shutting off the far side of the hut began to shudder. The sticks soon started falling to the ground. The animals, he had told me, had to be allowed to chew their way out: then they would believe that the structure belonged to them. We waited for another minute, then a creature which contrived to look both utterly alien and perfectly matched to this place emerged from the hole it had made. The onlookers cheered. It raised its big blunt head and sniffed the air, peering dimly towards the source of the noise. Then it waddled forward as I would expect an ankylosaur to have moved: hunched and heavy, dragging its belly and tail over the marshy ground.

It slipped into the pond, pushed its way through the waterweed and, suddenly slick and graceful, began to swim. Its head and back looked almost perfectly flat, emerging just an inch above the water, interrupted only by the little round ears. Half seal, half hippo, it paddled about in a circle. Then one of the cameramen shifted to get a better view and it flipped over, gave the surface of the pond a great crack with its tail and disappeared under the water. It emerged a moment later and began to swim along the bank, sniffing and poking its heavy snout into the rushes. The other one followed it into the pond, cutting a new path through the weed, occasionally displaying its fat rump as it dived, smooth and round as a dolphin.

This, as far as I can discover, was the first concrete step taken in Wales towards reintroducing an extinct mammal. Here, at Blaenein-ion, at the source of the stream which runs through the stunning Cwm Einion into the Dyfi estuary, a group of volunteers had enclosed three acres of land around an old carp pond. People had been talking about returning the beaver to Wales for years. Now, at last, something was happening.

It is not clear when beavers last lived in Britain, but they might have persisted until the mid-eighteenth century.[12] They were hunted to extinction for their beautiful warm fur and for castoreum, the secretion from the scent sacs close to the tail, which was used for making perfume and medicines. They once lived throughout our river systems, as much a part of our native ecosystem as they are in Canada today. Beverley in Yorkshire, Beverston in Gloucestershire, Barbon in Cumbria and Beverley Brook, which enters the River Thames at Battersea, are among the places named after them.[13] They are mild, plant-eating animals, popular with the people of the United Kingdom: an opinion poll found that 86 per cent were in favour of the beaver's reintroduction.[14] But listening to the small but powerful group of landowners fighting to prevent their reinstatement in this country, you could mistake the species in question for a sabre-toothed cat or velociraptor.

The body in charge of conservation in Scotland, Scottish Natural Heritage, started to investigate the idea of reintroducing beavers in 1994.[15] Landowners responded furiously. After ten years in which half a million pounds was spent assessing every possible danger the beavers might present, the Scottish government gave up and cancelled the project. An ecologist who was involved in this fiasco told me that, during a meeting which took place after six years of negotiations, one of the men who own the fishing rights on Scotland's rivers exclaimed: 'I hear what you say, and I can understand why some people like these animals, but I will not have them coming into my river and eating my fish.'

There was a deathly silence as the biologists realized that, through all those years of diplomacy and explanation, he still had not accepted that beavers are herbivorous.

Though beavers have been introduced, from 1924 onwards, to twenty-four other European countries without mishap,[16] and though they live among greater concentrations of salmon and other fish in Canada and Norway than we are blessed with in Britain, the landowners argued that they would stop the salmon migrating up the rivers, destroy their spawning beds and spread disease. At last, when every possible objection had been addressed from every possible angle, in 2009 eleven beavers were experimentally released into the Knapdale Forest in Argyll. The place into which they were reintroduced is unusual for its absence of salmon rivers – and, for that matter, of ideal beaver habitat.

By then, however, a number of beavers had 'escaped' from a wildlife park in Perthshire (it is widely believed that someone assisted their departure) and various other places, and had established themselves in the catchment of the River Tay, a famous and very expensive salmon river. As I write, the beavers (unlike those in the Knapdale Forest) are thriving and breeding freely, and the police and conservation authorities (the same conservation authorities who oversaw their release at Knapdale) are trying to catch them. 'They are being recaptured because their presence in the wild is illegal and because their welfare may be at risk,' Scottish Natural Heritage explains.[17] The illegal animals do not appear to have caused any harm, or any conflict with fishing interests, however. The accidental release could be seen as a more germane experiment than the official one.

The Scottish experience appears to have done nothing to reassure landowners in Wales. The Farmers' Union of Wales angrily denounced the work at Blaeneinion. It described the beavers as a 'non-native species', compared them to grey squirrels and claimed that they will spread diseases to their livestock.[18]

There is no intention to release these animals. Both are female, so there is no danger of proliferation. The point of the experiment is to establish, for the 162nd time in Europe,* that beavers are not the animals mentioned in the Book of Revelations, breathing fire and brimstone and slaying the third part of men. Their impacts on the plant and animal life in the enclosure will be studied, and the results

* Since 1924, there have been 161 reintroductions of beavers in Europe.[19]

will inform a possible reintroduction elsewhere in Wales. The favoured spot is currently the River Teifi, where, in the twelfth century, they were last recorded. But, as the Scottish saga suggests, this will be a long, slow process.

I watched the resurrected beavers for an hour or so, as they explored their new home, their lovely dense coats, which had given them so much trouble in their earlier incarnation, trailing bubbles as they paddled around the pond. They were much larger than I had expected. Occasionally they came up from under the water into a weedbed, and lay, crowned and garlanded with pond plants, indistinguishable, had you not seen them move, from mossy logs. One of them nibbled experimentally at a willow twig. Occasionally they dragged themselves out of the pond and sat on the bank, gazing around myopically. Their fur fluffed up immediately as the water streamed off it.

A man from the local paper turned up late, muttering and grumbling. 'Is this where they're releasing the badgers?'

'They're not badgers, they're –'

'Bloody hell, look at the size of that otter!'

Already they looked as if they owned the pond, sculling round it proprietorially, cutting paths through the weed, familiarizing themselves with the grasses and trees on which they would feed. Hard to see among the weeds and rushes, perfectly adapted to this interleaving of land and water, they looked as if they had always been here; as if they had never left.

The beaver is one of several missing animals that have been described as keystone species. A keystone species is one that has a larger impact on its environment than its numbers alone would suggest. This impact creates the conditions which allow other species to live there.

European beavers, unlike the North American species, build only small dams, but the changes they make to the flow of rivers, the branches and twigs they drag into the water, the burrows they excavate, the shallow ditches they create as they forage on the land and their felling of some of the riverside trees transforms their surroundings. They create habitats for water voles, otters, ducks, frogs, fish and insects. In Wyoming, where admittedly the ponds they make are larger than those in Europe, streams where beavers live harbour seventy-five times as many waterbirds as those without.[20]

In both Sweden and Poland (where European beavers live), the trout in beaver ponds are on average larger than those in the other parts of the streams: the ponds provide them with habitats and shelter they cannot find elsewhere.[21] Young salmon grow faster and are in better condition where beavers make their dams than in other stretches.[22] The total weight of all the creatures living in the water may be between two and five times greater in beaver ponds than in the undammed sections.[23] In Poland, beavers increase the number of bats hunting around the rivers, both because the population of flying insects increases as a result of their dams and their creation of swampy ground, and because they make gaps among the riverside trees in which the bats can hunt.[24] The trees they eat tend to be those which coppice or sucker well, such as aspen, willow and ash. The scrub this creates beside the rivers provides shelter for birds and mammals.

Our rivers, like the land, have suffered from intensive management. They have been straightened and canalized, dredged and cleared. The results have hurt both wildlife and people: by reducing the amount of time that water takes to flow from the tributaries into the lower reaches, we have ensured that the rivers are more likely to flood.

These policies often appear to have been informed by the same impulse that has driven some farmers to destroy lone trees and archaeological traces: a desire for tidiness. In the catchment of the River Wye, for example, the authorities spent large amounts of public money until the late 1990s on the pointless task of dragging what they called 'timber blockages' out of the tributaries. These great nests of branches took hundreds of years to accumulate. They were the prime habitat for a wide range of species, including the young of the salmon for which the river is renowned. Four hundred logjams were destroyed before someone realized that the policy resulted in nothing but harm.[25] The programme is likely to have helped cause the continuing fall in salmon numbers and the continuing rise in the number and intensity of the floods plaguing the towns around the river's lower course.[26]

Now the policy is being reversed. 'Let sleeping logs lie', the Wildlife Trusts advise.[27] They point out that woody debris in rivers helps to stabilize their banks and beds, that it traps sediments and provides shelter and food for insects and small animals, crayfish, fish, water voles, otters and birds.

In Yorkshire, where the town of Pickering has been flooded four times since 1999, to the great distress of its people, government agencies are now pulling woody debris back into the streams feeding Pickering Beck, in order to slow their flow.[28] This requires a good deal of labour and expense. There is a cheaper way of achieving the same result: releasing beavers. They drag branches into the water both to build dams and to create a food supply in the winter. They would keep protecting the town long after the funding for human workers ran out.

Beavers radically change the behaviour of a river. They slow it down. They reduce scouring and erosion. They trap much of the load it carries,[29] ensuring that the water runs more clearly. They create small wetlands and boggy areas. They make it more structurally diverse, providing homes for many other species. Far from spreading disease, as the Farmers' Union of Wales has claimed,[30] they could reduce it, as their dams filter out the sediments containing faecal bacteria.[31]

The more we understand about how ecosystems work, the less appropriate certain conservation policies appear. As I have explored the powerful effects that some species exert on animals and plants to which, at first, they have no obvious connection, I have begun to understand the extent to which the farmed and managed systems many conservationists defend are empty shells. They have lost not only their physical structure – the trees, shrubs and dead wood which provide habitats for so many species – but also many of the connections between the species which build an ecosystem. Most of the strands of the web of life in these places have been broken.

At first I struggled to identify the scientific principles that might inform rewilding. To formulate principles you must know what outcome you are trying to achieve. But rewilding, unlike conservation, has no fixed objective: it is driven not by human management but by natural processes. There is no point at which it can be said to have arrived. Rewilding of the kind that interests me does not seek to control the natural world, to re-create a particular ecosystem or landscape, but – having brought back some of the missing species – to allow it to find its own way.

But then I was struck by a thought which now seems obvious. The process is the outcome. The main aim of rewilding is to restore to the greatest extent possible ecology's dynamic interactions. In other words, the scientific principle behind rewilding is restoring what ecologists call trophic diversity. Trophic means relating to food and feeding. Restoring trophic diversity means enhancing the number of opportunities for animals, plants and other creatures to feed on each other; to rebuild the broken strands in the web of life. It means expanding the web both vertically and horizontally, increasing the number of trophic levels (top predators, middle predators, plant eaters, plants, carrion and detritus feeders) and creating opportunities for the number and complexity of relationships at every level to rise.

One of the most fascinating discoveries in modern ecology is an abundance of trophic cascades. A trophic cascade occurs when the animals at the top of the food chain – the top predators – change the numbers not just of their prey, but also of species with which they have no direct connection. Their impacts cascade down the food chain, in some cases radically changing the ecosystem, the landscape and even the chemical composition of the soil and the atmosphere.

The best-known example is the dramatic change that followed the reintroduction of wolves to the Yellowstone National Park in the United States. Seventy years after they had been exterminated, wolves were released into the park in 1995. When they arrived, many of the streamsides and riversides were almost bare, closely cropped by the high population of red deer (which in North America, confusingly, are called elk*). But as soon as the wolves arrived, this began to change. It was not just that they sharply reduced the number of deer, but they also altered their prey's behaviour. The deer avoided the places – particularly the valleys and gorges – where they could be caught most easily.[32]

In some places, trees on the riverbanks, until then constantly suppressed by browsing, quintupled in height in just six years.[33] The trees shaded and cooled the water and provided cover for fish and other animals,[34] changing the wildlife community which lived there. More seedlings and saplings survived. The bare valleys began reverting to aspen, willow and cottonwood forest. One apparent result is that the

---

* The European red deer, *Cervus elaphus*, and the North American elk, *Cervus canadensis*, are so closely related that until 2004 they were believed to be the same species.

number of songbirds increased: among the resurgent trees a study has found higher populations of species such as the song sparrow, warbling vireo, yellow warbler and willow flycatcher.[35]

The regrowth of the bankside forests also appears to have allowed the populations of both beavers and bison to expand: beaver colonies rose from one to twelve between 1996 and 2009.[36] The beavers then trigger all the effects I have just mentioned, creating niches for otters, muskrats, fish, frogs and reptiles. The returning trees have also stabilized the banks of the streams, reducing the rate of erosion and the movement of channels, narrowing the width of the streams and creating a greater diversity of pools and riffles.[37] Similar effects have been recorded in Zion National Park in Utah: where cougars are abundant, the streamsides are stable and fish numbers are high, where they are scarce, the rivers wander and fish numbers are three times lower.[38] The soil on the hillsides in Yellowstone, depleted through sheet erosion after the wolves were all killed and deer numbers rose, may now begin to build up again.[39] Conversely, on the grasslands where the deer and pronghorn antelope grazed heavily when their predators were absent, five years after the wolves returned, nitrogen in the soil declined by between a quarter and a half. This is because less of it is now recycled through dung.[40] This will change the species of plants that grow there and their numbers.

By hunting coyotes, the wolves allow the populations of smaller mammals – such as rabbits and mice – to rise, providing prey for hawks, weasels, foxes and badgers. Scavenging animals such as bald eagles and ravens feed on the remains of the deer the wolves kill. The return of the wolf appears to have increased the number of bears. They eat both the carrion abandoned by the wolves and the berries growing on the shrubs that have sprung back as the deer declined.[41] The bears also kill deer calves, reinforcing the impact of the wolves. The reintroduction of wolves to Yellowstone shows that a single species, allowed to pursue its natural behaviour, transforms almost every aspect of the ecosystem, and even alters the physical geography of the site, changing the shape and flow of the rivers and the erosion rates of the land.

There is no substitute for these complex relationships. Throughout the period in which wolves were absent from Yellowstone National Park, its managers tried to control the deer and contain their impacts – and failed.[42] Despite intense hunting and culling, willow trees disappeared from the meadows and aspens were in danger of vanishing from large areas of the park.[43] Even when hunting by humans is intense, its effects are likely to differ sharply from those of hunting by wolves. Wolves hunt at all times of the day and night, throughout the year. They pursue their prey, rather than killing it from a distance.[44] Wolves and humans hunt in different places and select different animals from the herd. Fencing might keep out the deer, but unlike wolves it does so entirely, while also excluding other animals and reducing the connectedness of the ecosystem.

Where salmon run, the reintroduction of wolves in North America could trigger even wider effects. The wolves create habitats for both salmon and beavers, and the beavers create further habitats for salmon, potentially boosting their numbers. The salmon are caught by bears, otters, eagles and ospreys. Their carcasses are often dragged or carried onto land. The nutrients they contain are distributed in the animals' dung. One study suggests that between 15 and 18 per cent of the nitrogen in the leaves of spruce trees within 500 metres of a salmon stream comes from the sea: it was brought upriver in the bodies of the salmon.[45] Top predators and keystone species unwittingly re-engineer the environment, even down to the composition of the soil.

A starker example is provided by the Arctic foxes introduced by fur trappers to some of the Aleutian islands – the sickle-shaped chain across the northern Pacific between Alaska and Siberia – where they are not native. Those islands with Arctic foxes are covered in shrubby tundra, those without foxes are covered in grass.[46] By hunting seabirds, the foxes have ensured that sixty times less guano is brought to the islands. This means that there is three times less phosphate in the soil than where they are absent. As a result, they have changed the entire natural system.

Human hunters might have imposed a similar change in the great steppes of Beringia, the landmass incorporating eastern Siberia, Alaska and the area in between (now covered by the Bering Straits, but exposed during the last Ice Age). Perhaps 15,000 years ago, hunters using small stone blades moved into the region that had hitherto been occupied by people hunting with sharpened bones or antlers.[47] Gradually, they wiped out the mammoths, musk oxen, bison and horses that grazed the steppes.* (When the glaciers blocking their passage into the rest of the Americas melted, they went on to wreak even greater havoc in the New World.) The result, it appears, was that they helped turn the steppe grasslands into mossy tundra. Much of this land has remained that way ever since.

As the Russian scientist Sergey Zimov has shown, grasslands, especially in the far north, are sustained by the animals that feed on them. By grazing, they make the grass more productive (in the steppes it grows five times faster than it does when it is not mowed). They recycle the soil's nutrients through their dung. The grass dries out the soil and smothers moss and lichens.[49] When the animals disappear, the self-reinforcing process goes into reverse. The dead grass, flopping over the soil, insulates it, ensuring that it stays cold, reducing the further growth of grass and encouraging moss to take over. As the moss begins to dominate, the soil becomes wetter and colder – still more hostile to grass. If the animals return, their trampling quickly breaks up the fragile layer of moss and lichens, allowing the grass to dominate again within one or two years.[50] The grazers in this habitat, in other words, are keystone species, flipping the entire ecosystem from one state to another.

This suggests, incidentally, that large-scale rewilding of the tundra, which Zimov and others promote, while a fascinating prospect, could have a damaging consequence. Moss is such a good insulator that it prevents even the top layer of soil from thawing.[51] It helps to stabilize the permafrost, locking up the methane it contains. If the moss layer is broken up and grasses return, while this might greatly increase the productivity and trophic diversity of the region, it could accelerate the melting which threatens to release large quantities of a powerful greenhouse gas. This is a reminder that rewilding, like any change we contemplate, has costs. In some cases the costs may outweigh the benefits.

Hunting by humans might also have transformed the environment of Australia. Before people arrived on that continent, it teemed with

* Mammoths might have been made more susceptible to extinction through hunting by the simultaneous shrinkage of their habitat. One paper suggests that this caused a 90 per cent reduction in their geographical range between 42,000 and 6,000 years ago.[48]

monsters. Among them was a spiny anteater the size of a pig; a giant herbivore a bit like a wombat, which weighed two tonnes; a marsupial tapir as big as a horse; a ten-foot kangaroo; a marsupial lion with opposable thumbs and a stronger bite than any other known mammal, which could prop itself up on its tail in order to stand on its hindlegs and slash with its tremendous claws; a horned tortoise eight feet long; a monitor lizard much bigger than the Nile crocodile. Most of these species, alongside many other marvellous beasts, disappeared between 40,000 and 50,000 years ago. At roughly the same time, the dense rainforests which covered much of that continent began to be replaced with the grass and scrubby trees which populate much of the outback today.

Two debates have raged among ecologists. Were these shifts caused by natural climate change or by humans? If, as now seems probable, they were caused by humans, were the extinctions of the giant animals the result of hunting or of the destruction of their habitats? Research published in the journal *Science* strongly suggests that humans hunted the large animals to extinction, and that the disappearance of the large animals then caused the destruction of the rainforests.[52]

Analysing the pollen and charcoal in cores taken from an ancient lake bed, and using the fungus that grows on the dung of large herbivores to measure their abundance, the researchers showed that the shift from rainforest to dry forest took place some 10,000 years before the climate dried out. Both the mass extinction and the change in habitat happened while the climate was stable. They also showed that fire began raging through the rainforests around a century after the large mammal populations collapsed; and that grass and scrub replaced the forests two or three centuries later. When the giant herbivores disappeared, they suggest, the twigs and leaves that would otherwise have been browsed began to build up on the forest floor, creating a fuel supply that allowed wildfires to destroy the rainforests and catalyse the shift to grass and scrub. The herbivorous monsters of Australia, like the mammoths and musk oxen of Beringia, appear to have sustained the ecosystem they browsed.

One of the interesting implications of the discovery of widespread trophic cascades is that removing an animal from a system – especially a top predator – may have counterintuitive and destructive results. For example, in many parts of Africa, people have killed lions and leopards in the belief that this will enhance their chances of survival and (among early European hunters) boost the herds of game. But one result has been an explosion in the population of olive baboons. They inflict such damage on crops and livestock that children have to be taken out of school to fend them off.[53] They also transmit intestinal worms to the people whose land they enter,[54] and appear to have reduced populations of wild game by preying on the young animals. Similarly, when conservationists in Florida sought to protect sea turtles by culling the raccoons which eat their eggs, they found that it caused the opposite effect. More turtle eggs were lost, as the raccoons were no longer eating the ghost crabs which also preyed on them.[55]

Perhaps the strangest example of these unexpected effects is the apparent link between the decline of vultures and the spread of rabies in India. In a remarkably short period, vultures have almost become extinct there as an accidental result of the use of a livestock drug called diclofenac, which turns out to be deadly to them when they eat the carcasses. As the number of vultures has collapsed, the carrion they ate is consumed instead by feral dogs. Their population, despite intense efforts to control it, has risen sharply as the vultures have declined. Dog bites are the cause of 95 per cent of the deaths from rabies in India, and the rising population means that more people are likely to catch the disease.[56] The vultures were also likely to have helped control animal diseases such as brucellosis, tuberculosis and anthrax, by clearing up infected meat.

Trophic cascades might once have dominated most ecosystems. The old belief among ecologists that natural systems were controlled only from the bottom up – that the abundance of plants controls the abundance of plant eaters, which controls the abundance of meat eaters – arose from the fact that many of the systems they were studying had already been greatly changed by people, not least through the reduction or extinction of top predators. Much of the richness and complexity – the trophic diversity – of these foodwebs was lost before it was recorded. We live in a shadowland, a dim, flattened relic of what there once was, of what there could be again.

7

Bring Back the Wolf

The fells contract, regroup in starker forms;
Dusk tightens on them, as the wind gets up
And stretches hungrily: tensed at the nape,
The coarse heath bristles like a living pelt.

William Dunlop
Landscape as Werewolf

We associate elephants, rhinos, lions and hyenas with the tropics. But until very recently (in geological terms) they lived in climates much colder than north-western Europe is today. Until around 40,000 years ago, the straight-tusked elephant (*Elephas antiquus*), closely related to the Asian elephant, roamed across much of Europe.[1] The woolly mammoth, which had an entirely different ecology, grazing on cold steppes (rather than browsing, like the straight-tusked elephant, in temperate forests) lasted longer: one relict population, isolated from human hunters in the fastness of Wrangel Island off the north coast of Siberia, survived until the Bronze Age.[2]

Three species of rhinoceros – the woolly, the Merck's and the narrow-nosed – lived in Europe at the same time as humans. Until roughly 40,000 years ago, Russia was haunted by two monstrous beasts, *Elasmotherium sibiricum* and *Elasmotherium caucasicum*. They were humpbacked rhinos the size of elephants, eight feet to the crest, weighing perhaps five tonnes. Elephants roamed across Europe, Asia, Africa and the Americas; rhinos never populated the Americas, but they lived throughout the Old World. Across the past 50,000 years the range and variety of these species have shrunk as humans have hunted them. They were exterminated first in Europe; then (in the case of elephants) in the Americas; then in the Middle East and North Africa; then in most of Asia; eventually in most of Africa. The animals conservationists are now desperately trying – and often failing – to save are the last, tiny populations of creatures which once dominated most of the earth's surface, so recently that we can almost stretch out our fingers and touch them.

When Trafalgar Square was excavated in the nineteenth century, presumably to build Nelson's column, the river gravels the builders exposed were found to be crammed with hippopotamus bones; these beasts wallowed, a little over 100,000 years ago, where tourists and pigeons cluster today. The same excavations – and those conducted in the square in the twentieth century – also revealed the bones of straight-tusked elephants, giant deer, giant aurochs and lions.[3] Lions raised their heads where the monument now stands long before Sir Edwin Landseer got to work.

They were larger than those now living in Africa but probably members of the same species. They hunted reindeer across the frozen wastes of Europe,[4] and survived in Britain until 11,000 years ago:[5] the beginning of the Mesolithic, when humans returned to the land after their long absence. Spotted hyenas (also still living in Africa) survived in Europe until roughly the same time[6] (their fossilized faeces have been found in Trafalgar Square[7]). Scimitar cats (*Homotherium* species), lion-sized with great curved fangs, preyed perhaps exclusively on young elephants and rhinos. These species – elephants and rhinos and the cats which ate them – are likely to have dominated the ecosystem during the previous inter-glacial period, which ended around 115,000 years ago (a blink of an eye in geological terms). The curious features that some of our plants possess may be ghostly adaptations to the way they fed.*

Elephants' habit of snapping or uprooting trees could explain why species such as oak, ash, beech, lime, sycamore, field maple, sweet chestnut, hazel, alder and willow can regrow from the point at which the stem is broken.† In eastern and southern Africa there are dozens of

---

* The idea behind these speculations was seeded in my mind by the forester Adam Thorogood.
† The only mention I have been able to find is in a paper by Oliver Rackham.[8] Coppicing and pollarding (resprouting at ground level, or from a cutting point higher

tree species which resprout – or coppice – from the snapped trunk, and ecologists recognize this as an evolutionary response to attacks by elephants.[9] By breaking African trees such as *mopane* or knobthorn acacia, elephants improve their food supply, as the shoots the damaged trees produce are easier to reach and more nutritious than older branches.[10] Trees that can survive the attention of elephants often come to dominate the places in which the animals live: the ability to coppice confers powerful selective advantages.

But somehow the obvious link – between coppicing and elephants – appears to have been missed by people studying European ecosystems. It is another example of Shifting Baseline Syndrome. Ecologists are not always aware of the extent to which the systems they study have been altered by humans: that the life they describe has been greatly simplified and diminished.

Elephants could also explain why understorey trees in Europe, such as holly, yew and box, are so resistant to breakage and have such strong roots, though they carry less weight than canopy trees and are subject to lower shear forces from the wind. They have to be tough, as they take much longer to become massive enough to withstand toppling or for their branches to grow out of the reach of trunks and tusks. The ability of some trees to survive the removal of much of their bark could be another adaptation: elephants often strip bark with their tusks. Elephant-proofing could account for the birch tree's pied coat: the black fissures make the white skin harder to strip cleanly.

The same evolutionary history could explain why traditional hedging, which relies on twisting, splintering and almost severing the living wood, is possible: the trees we use to make hedges would have had to survive similar attacks by elephants. Blackthorn, which possesses very long spines, seems over-engineered to deter browsing by deer; but not, perhaps, to deter browsing by rhinoceros.

---

up the trunk), Rackham says, are perhaps 'adaptations to recovering from the assaults of elephants and other giant herbivores. The extermination of the great tree-breaking beasts in Paleolithic times may have been mankind's first and farthest reaching influence on the world's forests.'

These animals,* with the trees they ate, were driven south by the last advance of the ice. By the time the ice retreated, they had been hunted to extinction. The trees returned to northern Europe, without the creatures they had evolved to resist. Our ecosystems are the spectral relics of another age, which, on evolution's timescale, is still close. The trees continue to arm themselves against threats which no longer exist, just as we still possess the psychological armoury required to live among monsters.

Even if these speculations do not lead to the reintroduction of elephants and rhinos, do they not render the commonplace astonishing? The notion that our most familiar trees are elephant-adapted, that we can see in their shadows the great beasts with which humans evolved, that the mark of these animals can be found in every park and avenue and leafy street, infuses the world with new wonders. Palaeoecology – the study of past ecosystems, crucial to an understanding of our own – feels like a portal through which we may pass into an enchanted kingdom.

They heard us coming long before I saw them, and the woods were now filled with strange sounds – yelping, roaring, whickering and a noise so deep that I heard it not only with my ears but also with my chest: a sustained, resonant drone, like the lowest note of a church organ. As we came within sight of the enclosure, the sounds intensified. The animals clustered around the gate. Thick-thighed, with small pert ankles and hooves, they looked like fat ladies in high heels. The rectangular blocky bodies were covered in dense bristles; their winter coats were almost blond. The delicate snouts were so long that they looked like little trunks. As the smell of the bucket reached her nostrils, the dominant female, crested and humped, a deep-bodied battering ram, barged the other beasts out of the way.

When the pellets were scattered on the ground, the boar purred and growled, occasionally exploding into shrieks and squeals as the big sow drove the others off the food. They ploughed up the soft soil, using not their little bleary eyes to find the food, but the sharper

---

* The straight-tusked elephant and the Merck's and narrow-nosed rhinoceros. Woolly mammoths and woolly rhinos, which were mostly grass eaters, living in cold dry steppes without trees, moved in with the cold weather.

organs in their snouts. Close to the fence the earth was churned and gouged; throughout the twelve hectares of the enclosure there were ruffles and furrows in the ground. This was why the boar had been brought here: to grub out the rhizomes of the bracken, which prevent tree seedlings from reaching the light, and to disturb the soil so that seeds could germinate. Though the remaining trees, now ancient, rained seed upon the ground here, none survived, because the bracken, released by heavy grazing from competition, had swarmed the bared land beneath them, creating an impenetrable barrier.

I would struggle to describe these boar as wild: the Dangerous Wild Animals Act forces their owners to act as zookeepers. The boar, like the beavers I saw in Wales, live behind high fences and electric charges. But elsewhere in Britain, they are starting to re-establish themselves, without permission from the authorities. The first major escape from boar farms here took place during the great gales of 1987, when trees crashed down on the fences. Since then they have continued to escape from farms and collections, and they have now founded at least four small colonies in southern England and possibly a fifth in western Scotland. They breed quickly. The government says that unless determined efforts are made to exterminate them, they will become established through much of England within twenty or thirty years.[11] It is a prospect that delights me, though I accept that not everyone shares this view.

Their reputation for ferocity has, like that of many large wild animals, been greatly exaggerated. It is true that they will attack dogs that chase them or people who corner them, but researchers who investigated this question concluded that, though they live throughout continental Europe, 'we have been unable to find any confirmed reports in the literature of wild boar making unprovoked attacks on humans'.[12] The government believes that the chances that they could transmit exotic diseases such as swine fever or foot and mouth to livestock are low, but they will cause damage to crops. This, it says, 'is likely to be small in comparison to agricultural damage from more common wildlife such as rabbits'.[13] They can also break into pig pens, kill the domestic boars and impregnate the sows.

On the other hand, the boar will catalyse some of the dynamic processes missing from our ecosystem. They are another keystone species, shaking up the places in which they live. The British woodland floor is peculiar in that it is often dominated by a single species, such as dog's mercury, wild garlic, bluebells, bracken, hart's tongue, male fern or brambles. These monocultures, like fields of wheat or rapeseed, may in some cases be the result of human intervention, such as the extirpation of the boar. To visit the Białowieża Forest in eastern Poland, which is as close to being an undisturbed ecosystem as any remaining in Europe, in May, when dozens of flower species jostle each other in an explosion of colour, is to see how much Britain is missing, and the extent to which boar transform their environment.

I understand people's concerns about the loss of those uninterrupted carpets of bluebells that have made some British woods famous. They are, I agree, stunning, just as fields of lavender or flax are stunning, but to me they are an indication not of the wealth of the ecosystem but of its poverty. One of the reasons why bluebells have been able to crowd out other species in the woods in which they grow is because the animal which previously kept them in check no longer roams there. Wild boar and bluebells live happily together, but perhaps not wild boar and only bluebells. By rooting and grubbing in the forest floor, by creating little ponds and miniature wetlands in their wallows, boar create habitats for a host of different plants and animals, a shifting mosaic of tiny ecological niches, opening and closing as the sounders pass through.[14] Boar are the untidiest animals to have lived in this country since the Ice Age. This should commend them to anyone with an interest in the natural world.

As the boar I watched were demonstrating, they allow trees to grow in places currently hostile to them. Another experiment, more advanced than this one, had revealed that where boar are allowed to root, both pine and birch seedlings establish themselves freely, whereas in the brakes without boar there is scarcely any regeneration.[15] In the enclosure I visited, the researchers had noticed that robins and dunnocks follow the boar around, feeding where they have overturned the ground. It could be that the robin evolved alongside the boar, rather as the oxpecker has evolved alongside large mammals in Africa, and that in the absence of boar it has now adopted human gardeners, who provide the same service.

The British government has washed its hands of the decision for which it should be responsible: what, if anything, to do about the returning boar. It has given landowners, both public and private, the task of deciding whether they should live or die.[16] This is a cop-out. The boar belong to everyone and no one, and we should be allowed to make a collective decision about what happens to them. It also ensures that, in most cases, the boar will be culled, without consultation, deliberation or research, because landlords are the group typically most hostile to the existence of any wild animals, except those they wish to hunt for sport. Already, boar are being killed here by the Forestry Commission and other owners at rates that could wipe them out. Among the commission's justifications is that they cause 'substantial damage' to woodlands.[17] What does this mean? The notion of damage to native ecosystems by a native species at numbers well below its natural population is nonsensical. What the Forestry Commission calls damage a biologist calls natural processes.

There might be a means of allaying the hostility even of the most resistant owners: allowing boar to become the one kind of animal they value – game. In Sweden, France, Germany, Poland and Italy, a powerful lobby now defends the boar out of self-interest. These are the hunters who stalk them in the woods and shoot them with high-powered rifles. Their licence fees are used to compensate the farmers whose crops the boars damage.[18] Licensed hunting in France appears to have transformed the public perception of this species, from agricultural pest to treasured native wildlife. And there are other, less destructive means of making money from them. Jenny Farrant, a farmer in East Sussex, first became aware of the wild boar on her land when they rooted up her hop bines.[19] Instead of waging war on them, however, she decided to make use of them, and now sells boar-watching holidays.[20] If the landowners now killing them indiscriminately give us the chance, we will soon come to value and cherish wild boar, just as we might come to value and cherish most of our once and future wildlife.

The boar I had come to see are one component of the most ambitious rewilding project in Britain. They live on an estate of 10,000 acres in the Scottish Highlands, purchased a few years ago from the family of a deceased Italian big-game hunter by an organization called Trees for Life. This estate, it hopes, will become the core of a great tract of rewilded land. The project is driven by one of the most singular men I have met.

Had someone described Alan Watson Featherstone to me and some of the beliefs he holds, I might never have written to him. Over the years, perhaps because I have spent too long in protest camps, I have developed a number of prejudices, which until now appeared to be rational: against people who believe in the significance of coincidences; against people who maintain that plants grow better if you love them; against people who live at the Findhorn Foundation (the spiritual community on the Moray Firth founded in the 1960s which, when I first visited it many years ago, seemed to be a permanent festival of fuzzy thinking and mumbo-jumbo); against men with ponytails. Alan belongs to all of these categories, yet he resembles none of the stereotypes I have, perhaps unfairly, constructed around such traits.

In the days that I spent with him roaming the glens and bens, and the nights staying in his tiny, beautiful eco-house in Findhorn, the moments accidentally eavesdropping on his video conferences and planning meetings, watching him organize, despite his opposing views, the talk I gave to the Foundation on the benefits of nuclear power, my prejudices fell apart. Efficient, entrepreneurial, focused, driven – this was a man who could have succeeded in any field. Without ever raising his voice or asserting himself, he transacted a vast amount of business, handling everyone he spoke to firmly and fluently. Fund-raising, recruitment, restructuring, redundancies, logistics, science, fieldwork – he appeared, without ever breaking sweat, to be on top of it all, yet he delegated tasks with no sign of territorialism or the other pathologies of 'Founder's Syndrome'.

I am used to being disappointed by visionaries, who often turn out to be lunatics or frauds, or to be afflicted with ossifying pride. But in this case, the more I listened, the more my respect grew. Never did I hear him hesitate or stumble. Every word was well chosen, every idea he expressed intelligible. He spoke softly and thoughtfully, engaging with the issues I raised, receptive to challenge or contradiction. He had a remarkable ability both to grasp the complexity and to keep his explanations simple, as if he had already condensed and summarized every subject I introduced. His is one of the most engaging minds I have come across.

Alan is a small man with delicate features and huge milky-blue eyes. He has a broad white beard and white hair which he wears, yes, in a ponytail. His movements are quick and busy; he springs over the

97

hillsides like a goat. He was born in Airdrie, a small industrial town close to Glasgow. When his family moved to Stirling, he began to take an interest in the woods and water that surrounded his house. When he left university, he travelled to North America, where he worked for four years as a tobacco farm labourer, a housepainter and a mining surveyor. The surveying work took him to remote places, where bears and moose were common sights.

'It was a transformative experience. It kicked off such a lot of wonder in me, and a desire to know those things. But I was working for the destruction of the earth. It was contrary to what my heart was saying was important.'

When he returned to Scotland, he went to live at Findhorn, where he worked in the foundation's gardens, coming to believe something I find hard to accept: that 'plants flourish in an atmosphere of love'. He visited Glen Affric, where some of the last remnants of the ancient Caledonian Forest grow, and was astonished by what he saw.

'I had never known that anything like this existed in Scotland. It looks like Canada or the western US. I had thought heather-covered hills and empty glens were natural. But I also realized that the Caledonian Forest remnants there were dying on their feet. I had a feeling in my gut: this land is calling out for help. Calling out to us. The feeling was there with me for years.'

In 1986, he organized an environmental conference in Findhorn, at the end of which people were asked 'to stand up and make a commitment to the earth'. He announced that he would launch a project to restore the Caledonian Forest. 'There was no going back then. I had no background, no experience, no qualifications. My degree is in electronics. But my passion was there. That's where the drive came from. The commitment to make it happen.'

At first he worked through the Findhorn Foundation; in 1989 he set up Trees for Life. He began by persuading some of the owners of estates on the north side of the Great Glen, the neat diagonal slash almost cutting Scotland in two, to let him plant or protect young trees on their land. He also began to recruit scientists to work alongside the project and to mobilize a volunteer army of mappers and planters. He started to form an astonishing plan.

Alan intends to reforest an area of some 1,000 square miles (roughly

10 per cent of the Highlands[21]) to the west of Inverness, encompassing glens Shiel, Moriston, Affric, Cannich, Strathfarrar, Orrin, Strathconon and Carron.[22] This area, which is mostly uninhabited, contains three of the largest remnants of the Caledonian Forest. His aim was to allow the existing forests to regenerate, to fill in the gaps through planting and to remove the exotic trees – Sitka spruce, lodgepole pine, Douglas fir, western hemlock – introduced for commercial forestry. The region would become a contiguous native forest, in which missing animal species could be reinstated and through which they could freely move, creating what he called 'the wild heart of the Highlands'. Within this area the trees would not be cut. The land, once they had become established, would not be managed. When I visited him, the volunteers working with Trees for Life were soon to plant their millionth tree.

To accelerate the project, Alan had set out to raise the money to buy an estate which could be solely devoted to rewilding. There are, in the Highlands, plenty of opportunities. The tragic history of this region – the Clearances that followed the Battle of Culloden (which took place not far from Findhorn) – has left most of the north of Scotland in the hands of a tiny number of landowners, few of whom live on their estates, and most of whom are not Scottish. In some places, making use of the right-to-buy laws passed by the Scottish parliament,[23] communities of smallholders have begun to regain a footing on the land. Some of these communities are rewilding parts of the land they have bought.

But in the rocky mountain core of the Highlands, where the soil is poor, the facilities sparse and most of the estates too large for communities to handle, human beings are an endangered species. It is one of the least-habited places in Europe, and people are unlikely ever to return in large numbers. Rewilding here, by contrast to some other promising places, conflicts with few people's aspirations.

As the new millennium began, Alan applied for grants, badgered philanthropists, boosted the membership, sold diaries and calendars and charged tourists and students to plant trees. He managed, by 2006, to raise £1.65 million, enough to buy the 10,000-acre Dundreggan estate in Glenmoriston.

The Italian owner had died intestate, and the sale of his property was tortuous. As so many of the absentee landlords of Scotland do, he had channelled his assets through holding companies in a tax haven:

in this case Liechtenstein. The legal knots took two years to untie. But, as Alan says, 'when you have a 250-year vision, you have to learn to relax a bit'.

Like most of the land in this region, the estate (with the exception of two small corners under forestry and sheep) was used for deer stalking. For a few weeks a year, a handful of people dressed in tweeds and brogues, steeped in Balmorality (see p. 149), travelled to Dundreggan to shoot stags. Otherwise, with the exception of the stalker (the deer manager), it was almost unvisited. But like the high sheep pastures in Wales the land had been scoured, and the last scraps of native forest were slowly succumbing to senescence. Without predators, fed by the estates in the winter, culled only lightly, the population of red deer had exploded. It has more than doubled in the Highlands since 1965.

The great Caledonian Forest, which once covered much of the Highlands, has been reduced, by people, sheep and deer, to around 1 per cent of its greatest extent. In some of the places where trees still exist, the youngest are 150 years old. The oldest were growing before the Battle of Culloden, when the political changes that destroyed much of Scotland's remaining forest began.

I arrived at Dundreggan – *Dul Dreagain*, Dragon's Hollow – on a day of fleeting sunlight and black clouds. Successive fronts were rolling up the Great Glen, driving the classic mixed weather of early April into the surrounding braes. Alan led me through a forest of ancient juniper bushes, twisting and bulging into fantastic shapes like zoophytes in *The Garden of Earthly Delights*.

After we had seen the boar, we walked through the old birches to the rocky ridge on which the last pines of the estate grew. Reduced – first by the shipwrights who logged the forests here, then by old age to a few hooked crones hunched over the hillside – these trees, which had clung to the rocks for a quarter of a millennium, were reaching the end of their lives. Great ginger branches had begun to shear off the trunks, tearing holes in their wide crowns. Young rowans grew high in the forks, sown there by birds. They too were among the last of their kind, as only those out of reach of the deer survived. On the track Alan found a pine marten scat, glittering with the iridescent wingcases of beetles.

Beyond the trees the bracken gave way to low, deer-cropped heather and ling, bronzed by the winter. As we climbed, a cold rain began to spatter, soaking the pages of my notebook. My ballpoint now scored a dotted line on the page, more imprint than ink, ghost-written. Head down against the wind, I noticed the fruiting bodies of the tiny lichens on the moor. It was as if an enamellist with a fine brush had crept up the mountainside, ornamenting them minutely with a shocking deep orange. We stepped over crimson plush cushions of sphagnum moss, like the upholstery in an Indian restaurant.

As we reached the peak of Binnilidh Bheag, the lights came on. The land, dull brown and tan before, flashed into colour. The sunshine, cleaned by the rain, was laser-sharp, and the wetness of the land accentuated its tints. Little pools on the moor below us exploded into points of light. The pines through which we had walked flared up: green fire amid the cool mauve of the bare birches. Beyond them the meanders and oxbows of the River Moriston snaked mercury, bulging with light.

The sun clipped out the features, making a scrapbook of the land. Ribbons of low trees surged up the small burns. Whale-grey rocks breached from the waves of heather. Among the beetle tints of the moor, a tiny green field emerged, and the broken wall around it rose into view, delineated by shadow. I thought of the love with which that field had been raised, suckled from the barrow with dung, primped and petted with mattock and spade, through brutal winters and cruel, deceptive springs, clothed with kale and neeps and tatties, before the Clearances snatched its makers from the land.

Would the rewilding of a large tract of the Highlands inflict similar damage upon the lives of its few remaining inhabitants, depriving them of their remaining means of making a living? This is a question I was unable to answer until I read a report published by the Scottish Gamekeepers' Association, which set out to document what it called 'the economic importance of red deer to Scotland's rural economy'.[24] It succeeded in demonstrating the opposite.

After denouncing attempts by conservationists and two of the more imaginative estates (Glenfeshie and Mar Lodge) to reduce the number of deer and encourage the reforestation of glens and braes, the association explained that in areas dominated by large landholdings (such as the region in which Trees for Life is working) deer stalking is the main source of employment. Other opportunities in such places, it says, are 'very limited'. So it commissioned a survey to discover how many people are employed in the management and running of the deer business on these estates.

It took as its case study the county of Sutherland, a wide territory in the far north of Scotland, covering 5,200 square kilometres. Of this, the report reveals, 4,000 square kilometres are in the hands of estates, which number just eighty-one. In other words, three-quarters of one of the largest counties in Britain is owned by eighty-one families, or by their secretive trusts in tax havens. Across the ten it sampled, covering 780 square kilometres, it found 112 people in full-time equivalent employment.[25] That means that just one person is employed by the dominant industry for every seven square kilometres, an area five times the size of Hyde Park. The association's figures suggest to me that the absentee owners and their monocultures of deer prevent not only the ecological regeneration of the region but also the economic regeneration.

The report also revealed that the income generated by stalking on the estates throughout Sutherland is £1.6 million. This is a tiny sum when spread across 4,000 square kilometres. Their expenditure on deer management is £4.7 million. In other words, stalking can be sustained there only because the bankers or oil sheikhs or mining magnates who own the land burn money on their expensive pastime. Even the tiny numbers of people employed by deer stalking are reliant on the irrational spending of absentee landlords, which could be terminated at any time.

Compare these figures with a study from the Isle of Mull, which discovered that colonization by white-tailed sea eagles has brought £5 million a year into its economy and supports 110 full-time jobs.[26] Thousands of people now travel to the island to watch the chicks hatching and fledging from the eagle hide at Glen Seilisdeir or to take an eagle cruise on Loch Shiel.[27] The eagles now account for half the enquiries at the visitor desk of the island's main ferry terminal.[28] A study commissioned by the Scottish government calculates that wildlife tourism in Scotland is already worth £276 million a year.[29] Rewilding and the reintroduction of other missing species could greatly enhance this figure, generating many more jobs than deer-stalking does today.

Gamekeeping is one of the greatest threats to this source of employment. Already one of the reintroduced sea eagles has been killed, alongside many other birds of prey, by poisoned meat laid out, most probably by a gamekeeper.[30] By damaging the potential for wildlife tourism in Scotland, the deer and grouse industries could be destroying more employment than they generate. This is not to dismiss the gamekeepers' right and need to work. But it does suggest that more people could make a living if the land were put to another use. The skills and local knowledge of the gamekeepers would be in high demand as wildlife-watching became a more important industry.

The wind filled my mouth and sealed my ears. It roared inside my head and numbed my hands. I watched a new cloud mass rolling towards us, dark as fate. We set off over the moor on the far side of the hill. But for porcupine tufts of unpalatable grass, the earth had been shaved: the plants, like those on some of the sheep pastures of Wales, were just half an inch high. Water welled up around my boots with every step. We came down to a gash in the soggy moor, torn from the land by a small dark stream. It had exposed the stumps and trunks of great pines, buried in the peat but now eroding out of the hillside.

'They haven't been dated yet, but they're close to the surface of the moor, so they're likely to be recent. There would probably have been trees alive here 150 years ago. You can go to almost any glen in the Highlands and you will find the stumps of the vanished forest. It's a tree graveyard.'

Heading down the hill by another route, we were hit by a storm of rain and hail. Driven by the wind, it was so hard and cold that I could feel the inner contours of my skull, sounded out by the sonar probing of ice and water.

We found ourselves among denser heath. In the midst of the storm, Alan stopped to show me hard-bitten birch twigs emerging from the heather. The lichen that encrusted them testified that they were much older than their size suggested. The path down the brae took us into another corner of the remnant forest, where a few more crabbed pines clung to the same ridge. The squall passed suddenly and the sun slashed through the sky, almost violent, its intensity somehow heightened by the coldness of my skin, as if, frozen hard, I could no longer absorb the concussion of light. I wrote now with the pen wedged in the palm of my hand, as my fingers could no longer close round it.

Here the bark of the ancient birches was corrugated like the cracked surface of a lava flow. The old pines had slowly heaved great rocks out of the soil, and now clutched them in their exposed roots, dangling over the ridge on which they grew, as if they were about to hurl them into the valley. The twigs of the great oaks were so heavy with lichen that at first I thought they were in leaf.

Beside the path was a glittering black dome, perhaps a yard across and two feet high. When I looked closely I saw that it was covered in large shiny ants, swarming furiously. There were so many that I could not see the nest beneath them. They had polished black heads, tawny collars, swollen abdomens striped black and pewter. Alan told me that these were wood ants. They were absorbing energy from the sun through their dark bodies.

'They will bring the warmth back down into the nest. When the sun goes behind a cloud, they slow down. If it stays behind the clouds, they return to the nest. Wood ants are solar engineers. They always build their nests with the main slope facing south: you can use them to orient yourself. They need a mixed woodland to survive: pine needles for building their nests and birch or aspen for the aphids they milk.'

And there, close to the nest, were the pale green trunks of aspen, their bark pitted as if it had been blasted with shotguns. Like the other species of the old forest, they had aged without progeny for many years, but now the volunteers who worked on the estate had placed guards around the suckers the old trees threw up, in some places as far as fifty yards from their trunks. The suckers grew much faster than seedlings could, as they could draw upon the network of roots: even in this harsh land, the young shoots could rise by over a yard in ten weeks. The ants had already been seen tending and taxing the aphids which feed on the sap, extracting the honeydew they secrete.[31]

Aspen, favoured by deer, is now rare in the Highlands. Trees for Life had been mapping its remaining stands, protecting the suckers and cutting root sections to propagate and grow in places from which the tree is missing. Aspens support rare insects, lichen and fungi, but Alan also had another species in mind. The estate extends to the river, which looks like an excellent habitat for beavers. Like deer, they will feed on aspen in preference to any other plant; its suckering habit is likely to be an adaptation to the assault it encountered wherever it grew.

'We are getting the habitat ready. But we don't own enough of the river to do it all ourselves. We'll have to persuade the neighbouring landowners to help.'

In the autumn volunteers swarm the woods, collecting birch catkins. They return in the spring to find pine cones, and lay them out to crack in the sun. They pass the seed to the Forestry Commission, aware that local stock is likely to prosper here more readily than seedlings from elsewhere. Trees for Life had been propagating the less common species – aspen, juniper, holly, hazel, dwarf birch – in its own nursery. 'But that will probably have to go in the restructuring.'

Alan was continually, and unsentimentally, adjusting the operation to match its fluctuating budget. He appeared unabashed by these decisions.

In the other glens in which they worked, the Trees for Life volunteers were restoring alder carr, blocking drainage ditches to raise the water levels and replanting the missing trees. They were fencing areas where eared willow grows and planting hazel to create more habitat for red squirrels. Already in some of these places willow warblers had returned and water voles were spreading into new habitat.[32] They were creating a corridor of woodland which would, in time, connect Glenmoriston with Glen Affric, five miles to the north.

As Trees for Life reduces the number of deer through culling and draws them away by shifting the feeding stations, Alan explained, he expects birch to colonize much of the open ground, followed by pine, then oak, ash, wych elm, holly and hazel. The north-facing slopes were once dominated by pine; the lower southern slopes by broad-leaved trees. 'We don't expect trees to return everywhere here. It would be sparser in places: a mosaic. Not like the wall-to-wall conifers in the plantations.

'It was when I saw these places in the 1980s that I felt called to do something about it. Seeing the stumps in the peat and the remnant trees, I asked myself: what's the message in the land? What's the story it's telling us? My question was: "What's Nature seeking to do here?" That is crucially different from the ethos of human domination. Rewilding is about humility, about stepping back.'

This land, he hoped, would within fifty years be used by capercaillie, ospreys, golden eagles, red squirrels, boar, beavers, perhaps lynx. But these were the less contentious of his proposals. 'My aim is to have wolves back in Scotland by 2043. That would be 300 years after the last one is said to have been killed here. It's one generation from now. Ecologically they could live here today. The obstacles are cultural and economic.'

I stood, braced against the bitter wind, under the torn canopy of the old trees, absorbing what he had just said, my synapses firing, my thoughts slipping across a world that had suddenly become more labile, more thrilling, less predictable than any I had pictured until then. I felt a shiver of transgression – of sharing a thought forbidden, abhorred – mingled with confusion and doubt. Was this possible? Permissible? Even to imagine?

We ate our sandwiches in Alan's car, then put our seats back and slept. He fell asleep immediately, as if he had turned off a light. I drifted for a while. Wood ants swarmed over the land, darkening the earth, each one carrying a seed in its mandibles, now frantically ruffling through the earth with their snouts, eared and bristling, shoving in the seeds and scraping the soil back with their trotters, swarming on, tusks and antennae, over the mountains and through the next glen . . .

I will not try to disguise my reasons for wanting to see missing animals reintroduced. It is not, as the previous chapter might have suggested, the desire to control floods, or reduce erosion or hinder the spread of disease, though all these might be useful side-effects. My reasons arise from my delight in the marvels of nature, its richness and its limitless capacity to surprise; from the sense of freedom, of the thrill that comes from roaming in a landscape or seascape without knowing what I might see next, what might loom from the woods or water, what might be watching me without my knowledge. It is the sense that without these animals the ecosystem is lopsided, abridged, dysfunctional. I can produce reasons scientific, economic, historic and hygienic, but none of those describe my motivation.

Living in Britain, I am constantly reminded of the scale of our loss. According to the biologist David Hetherington, who runs the Cairngorms Wildcat Project, the United Kingdom is 'the largest country in Europe and almost the whole world' which no longer possesses any of its big carnivores.[33] It has also lost more of its large native species – both carnivores and herbivores – than any other European country except the Republic of Ireland. Britain also happens to be the slowest and most reluctant of any European nation to begin rewilding the land and reintroducing its missing species.

Perhaps this is connected to the fact that we have one of the highest concentrations of land ownership in the world.[34] Large landowners, who are often (though not universally) hostile towards any wild animals that might compete with or prey upon the animals they hunt, and often deeply suspicious of proposed changes to the way they manage their estates, are peculiarly powerful here. Though they and their views tend to belong to a very small minority, they dominate rural policy, and little can be done without their agreement.

A group called Rewilding Europe intends to catalyse the restoration of ecological processes across a million hectares of the Continent by 2020, and to encourage other bodies to take on a further 10 million.[35] It appears to be on schedule. In the first phase of restoration, it is working in the Danube delta, the southern and eastern Carpathians, the Velebit Mountains of Croatia and the *dehesa* (or *montado*) – the wooded savannahs – of Spain and Portugal.

The Danube delta contains the world's largest reedbeds and the last primeval forest in Romania, some of whose trees are 700 years old. Despite the best efforts of the former dictator, Nicolae Ceaușescu, and a wildly misconceived project by the World Bank, much of the marshland remains undrained, and many of its rivers still flow freely. Many of the dykes, agricultural schemes and pumping stations the developers commissioned have collapsed or ceased to function. Here there are pelicans, bitterns, eight species of heron, hobbies, red-footed falcons,

rollers and bee-eaters, waders, geese and grebes of many species, hoo-
poes, orioles, fire-bellied toads, giant catfish, sturgeon weighing
almost a tonne. But the native mammals have been hunted nearly to
extinction.

The great forests and floodplains of the eastern Carpathians, divided between Poland, Slovakia and Ukraine, still contain bison, lynx,
wolves, bears and beavers. As farmers have moved off the land, their
fragmented ecosystems are beginning to reconnect. In Poland over a
million people – most of them Polish – travel to these mountains every
year to walk and watch the animals. In Slovakia, however, the
old-growth forests are still being logged, as the potential for generating money by other means has not been fully grasped.

The southern Carpathians, in Romania, through which I once
walked and camped for three enchanted weeks, still possess in many
parts a natural treeline. The great beech forests of the valleys give way
to firs on the slopes, which diminish into scrub, then high alpine pastures, where, as the snow retreats, crocuses, saxifrages, pinks and
primroses spring up. The clearings in the lowland forest were, when I
visited, so thick with butterflies that it was sometimes hard to see the
path. There are wolves, boar and bears in these mountains, large parts
of which are already well protected. The rewilders want to reduce
hunting to raise the number of chamois and red deer, and to reintroduce bison, beavers and griffon vultures. In 2012, the first five bison,
which had been extinct in Romania for 160 years, were released into
the Vanatori Neamt reserve.[36]

The Velebit Mountains, which rise almost 6,000 feet from the Adriatic coast, already support lynx, wildcats, wolves, bears, chamois and
boar, as well as a magnificent variety of birds and snakes and butterflies. In the *dehesas* and *montados* of Spain and Portugal, the Iberian
lynx, extinct across much of its former range and now the world's
most endangered wildcat, is slowly recovering, through the reintroduction of animals bred in zoos. The governments of the two countries
have set aside over a million hectares of this land for conservation, to
protect the lynx, the Spanish imperial eagles, the vultures, Iberian ibex
and other rare wildlife that lives there.

In each of these places, Rewilding Europe is seeking to demonstrate that restoring ecological processes makes more money for local people than was generated by the industries that formerly used the land. It is hoping to reintroduce missing species and to raise the populations of animals which until now have been persecuted. While talking to two of its officers I was told something I have not heard from environmentalists in a long time: 'money is not a problem'. Public enthusiasm for rewilding on the Continent is so great that their initial projects are fully funded.

In 1997, wildlife groups and travel companies formed the Pan Parks Foundation, which hopes to secure a further million hectares of self-willed land in Europe.* So far it has protected 240,000 hectares, in Sweden, Finland, Russia, Estonia, Lithuania and Belarus, Romania, Bulgaria, Italy and Portugal. In 2012, after ten years of negotiations, it created what it calls its first 'transboundary wilderness': a single protected area incorporating national parks in Finland and Russia in which no hunting, grazing, logging, mining or any other extractive industry is allowed.†

The conservation group WWF is helping to protect around a million hectares in the Carpathian Mountains and the Danube catchment, connecting existing national parks and rewilded lands in Serbia and Romania.[38] A coalition of wildlife groups called Wild Europe hopes to allow wildlife to move between protected areas all over the continent, by creating ecological corridors and restoring degraded land.[39] The Polish government intends to increase the wild land around the Białowieża Forest, the largest expanse of primeval forest in Europe.[40] The German government has now pledged to rewild 2 per cent of its land by 2020.[41]

Almost everywhere, except Britain and Ireland, large charismatic species are returning. Wolves have spread across most of Europe.

---

* It uses a definition of wilderness produced by a coalition of wildlife groups: 'Wilderness areas are large unmodified or only slightly modified natural areas, governed by natural processes, without human intervention, infrastructure or permanent habitation, which should be protected and overseen so as to preserve their natural condition and to offer people the opportunity to experience the spiritual quality of nature.'

† It connects the Oulanka and Paanajärvi national parks, creating a single 'wilderness' of 132,000 hectares.[37]

Between 1927 and 1993, the wolf was extinct in France. Now, helped only by the restraint of people who might otherwise have killed them, there are over 200 wolves there, in at least twenty packs, some of which have spilt into Switzerland.[42] The wolves which began to arrive in Germany from Poland in the late 1990s – almost a century after the species became extinct there – have now formed around a dozen packs.[43] Since they were almost exterminated in the 1970s, wolf numbers in Spain have quintupled, to around 2,500. They have also grown rapidly in Italy and Poland.[44] In 2011, 113 years after the species became extinct there, a camera trap in Belgium produced footage of a wolf dragging away the carcass of a deer.[45] Another one – or possibly the same one – was seen in the Netherlands in the same year.[46]

Bears on the Continent have more than doubled in number over the past forty years. Though they have declined to critically low levels in France, Italy and Spain, they have been allowed to multiply in Scandinavia, the Baltic states, eastern Europe, the Balkans and Russia. Now there are some 25,000 in Europe.[47] Extinct in Austria since the nineteenth century, they have slowly been reintroduced, though with a fair number of setbacks: they are the most difficult and dangerous of Europe's large wild animals.

The population of European lynx, reduced to almost nothing a century ago, began to recover a little in the 1950s; since 1970 it has more than tripled, to around 10,000.[48] During this period lynx have been reintroduced to the Jura Mountains and the Alps in Switzerland, to the Dinaric Mountains in Slovenia, the Bohemian Forest in the Czech Republic and the Harz Mountains in Germany. They have reintroduced themselves in other places.

The European bison, or wisent, the magnificent animal whose bulls can weigh over a tonne, once roamed the forests and steppes from central Russia to Spain. Soon after the end of the First World War it became extinct in the wild, and only 54 wisent remained alive in captivity.[49] Some of their descendants were released into the Białowieża Forest in eastern Poland in 1952. Soon after the collapse of Soviet communism, I spent a fortnight there in late spring, pedalling silently down the sandy paths on a hired bicycle, then stalking as quietly as I could through the trees whenever I came to a promising spot. Scarcely touched by foresters, this is an ecosystem of the kind which must have been familiar to the people of the early Mesolithic. Oak and lime trees with trunks twice as wide as the length of my bicycle rose perhaps 100 feet without branching. Where they had fallen they formed an unscalable barrier, which dammed the spongy ground, creating small pools. The forest floor was a maze of dead wood. Between the toppled trunks it frothed with ramsons, celandines, spring peas and may lilies. I disturbed boar with their piglets, red squirrels, hazel grouse, a huge bird that might have been an eagle owl, a black woodpecker. Hiding in the reeds beside a river that ran through the forest, waiting in vain for the beavers which had felled the birch trees with cartoon precision, I saw a great snipe fly overhead. Along the streams on the edge of the forest at night, every bush appeared to contain a nightingale. Black storks scoured the meadows, among a hubbub of frogs and corncrakes.

I saw the bison only twice. On the first occasion I walked around a curve in the path and met an animal which looked more like a Christian depiction of the Devil than any other creature I have seen. We both stopped. I was close enough to see the mucus in her tear ducts. She had small, hooked black horns which gleamed slightly in the soft light of the forest, heavy brows and eyes so dark that I could not distinguish the irises from the pupils. She wore a neat brown beard and an oddly human fringe between her horns. Her back rose to a crest then tapered away to a narrow rump, from which a black tail, slim as a whip, now twitched. She flared her nostrils and raised her chin. I fancied I could smell her sweet, beery breath. We watched each other for several minutes. I stayed so still that I could feel the blood pounding in my neck. Eventually she tossed her head, danced a couple of steps then turned, trotted back down the path and cantered away through the trees.

On the second occasion, I had hidden among some bushes overlooking a pond I had found deep in the forest, which was surrounded by spoor. I had waited for no more than an hour when I was struck by the impression that the trees were moving. I blinked and looked again: a large herd of wisent had materialized beside the water. It was hard to believe that animals of this size could have arrived so quietly. The cows drank while their fluffy calves stood beside them, their front legs in the water. The great slab-sided bulls burnt ginger in the spotlight of the pond's clearing. Now I could hear them snuffling the water, occasionally snorting and softly groaning. After perhaps twenty minutes, the forest began to move again as the bulls hauled their bulk from the pond then stood on the bank, looking around as the cows raised their heads from the water, beards dripping, before backing away through the mud, while the calves jostled, afraid that they would lose touch with their mothers. Wisent have now been reintroduced to many parts of eastern Europe, to Germany, Spain, the Netherlands and Denmark, though in some of these cases they remain within enclosures, awaiting a wider release. The population has risen to around 3,000, but, as they are all descendants of just thirteen animals, the genetic base is dangerously small.

Beavers have been released, at the latest count, on 161 occasions in Europe.[50] Reduced by 1900 to tiny populations on the Elbe, the Rhône, in the Telemark district of Norway and the Pripet marshes in Belarus, their numbers have risen 1,000-fold, to some 700,000.[51] Golden jackals, after being driven out of much of Europe, are now multiplying in Bulgaria, Hungary and the Balkans, and moving into parts of Italy and Austria from which they might have been absent since the Iron Age. (The date of their disappearance is quite speculative, as the fossil and historical evidence is patchy.)

But this ecological revolution, though occurring in almost every other country in Europe, has left Britain untouched. There are several reasons. Species such as wolves which can extend their range freely on the Continent cannot reach these islands unless someone buys them a ferry ticket. Farmers have been slower to leave the land here than they have elsewhere: it seems that the further people are from the towns, the sooner they give up, perhaps because of the sense that life elsewhere is passing them by. Few parts of Britain are as far from large settlements as some of the farmland in Spain and Portugal, southern France and central and eastern Europe.

But this explains only part of the difference. The contrast between attitudes to nature in these isles and on the Continent is striking. I have often been told that Britain is too small and crowded for rewilding, though the same consideration has not stopped the Netherlands, which has much less land suitable for cultivation. I have also been told that we cannot afford it; though this has not inhibited Romania or Bulgaria or Ukraine.

Perhaps Britain is the most zoophobic nation in Europe. We appear to possess a deep fear of wild animals, even those which can do us no possible harm. This could be because this was one of the first nations to become largely urbanized, or because much of the countryside is controlled by that small but peculiarly powerful class, which often seems to be antagonistic to any wildlife not classified as game. But it is also clear that, partly perhaps because of the popularity of wildlife programmes, enthusiasm for the idea of restoring our native wildlife is growing – everywhere except among the few thousand people who own most of the countryside. It is an unfortunate quirk of fate that those likely to exert the most influence over the question of whether or not our missing species are reintroduced are those who are most resistant to the idea. But in the case I am about to discuss, it is not only the landowners who are likely to voice strong objections.

The deadly ferocity of the wolf is a story to which we are exposed early and often. It swallows grandmothers then borrows their clothes. It dresses as a sheep or a sheep dog to pursue its wicked schemes. It blows down houses. It hybridizes with people to spread havoc through merely human society. Christianity equates wolves with evil and greed, though they played a more positive role in the foundation myths of some cultures, such as the Turkics, Chechens, Inuit and Romans.

To what extent are the horror stories true? Wolves have certainly killed people. A comprehensive review of recorded wolf attacks from 1557 until the present found that unprovoked attacks by non-rabid wolves are 'very rare', and that almost all of them took place prior to the twentieth century.[52] Researchers found that eight people have been injured by wolf attacks in Europe in the past twenty years, but no one has been killed. There are nearly 20,000 wolves in Europe. During the past fifty years, five people have been killed by rabid wolves on the Continent and four by wolves without rabies, four by each category in Russia (where there are 40,000 wolves) and none in North America (where there are 60,000). Wolves not carrying rabies are most likely to attack when they have lost their fear of humans and live among them, or when they have been cornered or trapped.

There is no rabies in Britain,* and any wolves brought here for

---

* Except among bats, which tend not to spread it to any other form of wildlife. (Vampire bats in South America are another matter: they spread rabies to other species,

reintroduction would be screened and quarantined. If wolves retain their fear of humans (which I will discuss in a moment), attacks are likely to be extremely rare, perhaps non-existent. The chance of being killed by a wolf in Europe, even where they are abundant, is much smaller than the chance of being struck by lightning, or of being slain by the wrong kind of bedroom slippers (the cause of a number of fatal plunges down stairs) or by a collapsing deckchair. Even so, their reintroduction is a risk, however small, that will be imposed on other people. So it should happen only with wide public consent.

We expect the people of other countries to conserve far more dangerous animals than wolves: lions, tigers, leopards, elephants, hippos, crocodiles and Cape buffalo, for example. Many people in rich nations give money to the wildlife groups protecting them. Are dangerous (or in this case not very dangerous) wild animals something we choose to impose on other people, but not upon ourselves?

Wolves do present a more realistic threat to livestock, especially sheep. For reasons which are not well understood, they prefer to hunt wild game, though sheep are easier to catch.[53] Even so, wherever they live they clash with livestock farmers. The impacts across the whole industry are small (less than 0.1 per cent of the sheep kept in the parts of America where wolves live are killed by them,[54] and 0.35 per cent in Italy[55]), but their effects on an individual farmer can be greater, especially if a local wolf has developed a taste for mutton. Occasionally a wolf will slaughter a large number of sheep in a single attack (wolves will return to their kill for weeks if there is enough meat: mass killing is an attempt to create a larder).

Across France, Greece, Italy, Austria, Spain and Portugal, an average of €2 million a year is paid out in compensation to farmers who have lost animals in wolf attacks, and roughly the same amount is spent on preventing them.[56] Though these figures are small, the agencies handing out the compensation money could be overpaying, as dog attacks are often blamed on wolves (in Italy, for example, there are 900,000 feral or free-ranging dogs and just 400 or 500 wolves[57]) and some claims are probably fraudulent.

---

including humans. Some of the goldminers in Roraima told me of terrifying outbreaks caused by vampires in areas they had prospected in the western Amazon.)

There is a possible deterrent which has not been widely discussed in Europe, though it is used in South Africa to protect animals against lions and other predators, and in America to tackle coyotes. The live-stock protection collar carries a chemical in two capsules at the animal's throat, ensuring that a predator ingests it when it kills. In the US, sheep farmers load it with deadly poisons, but an emetic (a com-pound which causes vomiting) could deter predators from attacking that kind of livestock again. A Swiss biologist has designed another clever device: a collar that monitors a sheep's heartbeat. If the rate rises and stays high for long enough, the collar sends a text message to the farmer. Sheep become distressed as soon as they see a wolf, so the farmer could have time to reach them before the wolves attack.[58] The same collar could also produce noises of the kind a human would make, to frighten the wolves away before the farmer arrived.

Alternatively, a wolf that makes a habit of killing sheep can simply be shot. Though I hate the thought of killing wolves, and could never do so myself, I think we should be able to love wildlife without being unreasonably sentimental.

In fact hunting, strange as this may sound, could be the wolf's sal-vation. There are three reasons for this. The first is that, as with wild boar, allowing licensed hunters to shoot wolves is likely to create a powerful lobby for their protection, just as anglers have become the staunchest defenders of fish stocks. The second is that it shows other people that the animals are under control. I feel we control our wild-life too much, but the wolf has a public relations problem, and the idea that it should be allowed to roam and breed without check is likely to be too much for many people to contemplate. Licensed hunt-ing in Sweden has gone some way towards making the wolf politically acceptable there, after it reintroduced itself from Finland in the 1970s, provoking widespread demands that it be exterminated.[59] I was told something similar by a forest officer in Slovenia: were it not for the authorized hunting of wolves and bears, they would be wiped out by unauthorized hunters, concerned that no one was managing them. In both countries, however, the number of wolves hunters are allowed to shoot every year is a highly contentious issue: over-hunting is sup-pressing the population of wolves to the extent that their genetic viability is threatened.

The third and most important reason is that it keeps the wolves afraid. As the review of wolf attacks suggests, the best means of protecting people from wolves is to ensure that wolves go nowhere near them. Nothing is likely to do this more effectively than an occasional shooting. The same tactic could be used to prevent wolves from migrating into areas in which they are not welcome. At other times people have hunted the wolf in order to eliminate it. Now we might hunt the wolf in order to preserve it (but not within protected areas).

The last British wolf is widely believed to have been killed in the Findhorn Valley, close to where Alan lives, in 1743, though the story is treated as apocryphal by the great rural historian Oliver Rackham. The last definite record of a wolf in Britain, he says, was the massive bounty paid for an animal killed in Sutherland in 1621.[60] Wolves survived for longer in many parts of the Continent until they were reduced, during the twentieth century, to remnant populations in Spain, Italy, Scandinavia and eastern Europe. Their return to much of Europe, which in many places has been greeted enthusiastically, is perhaps the clearest sign of a radical change in attitudes to nature over the past forty years or so, a change that has been taking place more slowly in Britain but which, even so, is tangible.

Wolves range widely and can live almost anywhere: tundra, deserts, forests, mountains, moorland, farmland, cities. When they are not killed, they quickly re-establish themselves. There is one part of Britain which has all the characteristics required for their reintroduction: the Scottish Highlands. There the population of red deer and roe deer is not only high enough to support them but far too high. The human population is far lower than in many parts of Europe (such as eastern Germany and the Apennines) in which wolves live today. There are few roads, which means that they are unlikely to be killed by cars. The Highlands could probably support around 250 wolves, which should be enough to keep the population viable.[61] England and Wales are less suitable, as they have fewer deer; in Wales deer have been almost obliterated.

While wolves and sheep may not be the perfect social mix, introducing wolves to Scotland's deer population could, one study suggests, benefit even the big estate owners.[62] The overpopulation of deer, while it pleases the stalkers, presents them with a major management problem. Suppressing the population to the extent recommended by the Deer Commission is a labour-intensive and expensive business. People pay to stalk and shoot the stags, but the profits tend to be offset by the losses incurred in shooting the hinds (the females), with the result that most estates either make a loss or just break even. The scientists who have modelled the effects of reintroducing wolves find that it is likely to make them more profitable. While wolves would reduce the number of stags, they would also avert the need for a hind cull. The result would be that the estates would make a profit of £800 a year for every ten square kilometres from deer keeping, rather than £550.[63] The remaining stags are also likely to get bigger, as there will be more food for each deer, which could mean that people would pay more to shoot them. The wolves, the model suggests, are likely to reduce the deer in the Highlands to around half their current number.

By killing and deterring deer, wolves allow woodland to regenerate. A study published in the *European Journal of Forest Research* suggests that hunting by humans is a less effective means of protecting forestry than hunting by wild predators.[64] Wolves not only suppress the population but radically alter the behaviour of the deer. They might also reduce the number of cases of Lyme disease, a debilitating and (in its advanced stages) sometimes incurable illness spread to humans by deer ticks.[65] While we are well aware of the wolf's unhelpful contribution to sheep farming, we are perhaps less aware that this will be partly balanced by their killing of foxes, which often carry off lambs. For the same reason, they are likely to be beneficial to grouse moors and pheasant shoots. In North America, most of the compensation paid to farmers for the damage done by wildlife takes the form of payments for crops eaten by deer, not for livestock eaten by wolves and coyotes.[66] It is possible, though I have not yet been able to find comparative figures, that wolves there could in fact increase the overall production of food for humans.

Again, it would be deceptive to claim that I would like to see wolves reintroduced because they kill foxes or reduce disease or assist the owners of grouse moors and deer estates. I want to see wolves reintroduced because wolves are fascinating, and because they help to reintroduce the complexity and trophic diversity in which our ecosystems are lacking. I want to see wolves reintroduced because they feel to me like the shadow that fleets between systole and diastole, because they are the necessary monsters of the mind, inhabitants of the more passionate world against which we have locked our doors. The return of the wolf also makes the introduction of other missing species – such as boar and moose – more viable, as their populations will be checked without the need for human intervention. But it should happen only if there is broad public enthusiasm for the project.

A survey conducted in Scotland suggests that people are less hostile to the reintroduction of the wolf than one might have imagined. The idea meets with slightly more favour than disfavour among rural people, and is welcomed a little more firmly by urban people.[67] Even sheep farmers, surprisingly, were split: antagonistic on balance, but not universally so. The researchers who conducted the survey suggest that this could be because they make most of their money from subsidies, rather than from selling lamb. Only the National Farmers' Union of Scotland was fiercely opposed, suggesting that, as in many other matters, it may not be representative of its members (farmers' unions in Britain tend to be dominated by large landowners with strongly conservative views). I wonder whether the Farmers' Union of Wales might have misrepresented the attitude of Welsh farmers towards beavers.

While the wolf is a hard sell, another large predator could be introduced today, at no risk to people and little risk even to sheep. The lynx, until recently, was assumed to have belonged only to prehistoric Britain, unknown to the people even of the Neolithic.[68] But recent finds have radically changed that assessment. First, lynx bones discovered in a cave in northern Scotland and two sites in north Yorkshire were dated at around 1,800 years old, dragging the species towards the present by some 4,000 years. Another cave in Yorkshire then produced a bone around 1,500 years old.[69] That is now the most recent fossil evidence, but the cultural evidence for their continued existence in Britain extends a little further.

Cumbric is a Celtic language similar to Welsh that was spoken in the north of England and southern Scotland – the territory, once much larger than the current county, known as Cumbria. A seventh-century Cumbric manuscript records the battles of *Hen Ogledd*, the Old North. Among these gory sagas sits, incongruously, a sad and beautiful nursery rhyme or lullaby. It is called *Pais Dinogad*: Dinogad's Shift. The mother tells her son, Dinogad, of his dead father's prowess as a hunter.

> Dinogad's shift is speckled, speckled,
> It was made from the pelts of martens . . .
> When your father went to the mountains
> He would bring back a roebuck, a boar, a stag,
> A speckled grouse from the mountain,
> And a fish from the Derwennydd falls.
> At whatever your father aimed his spear –
> Be it a boar, *llewyn*, or a fox –
> None would escape but that had strong wings.*

This is not, in other words, an account like the story of *Cath Palug* in the *Black Book of Carmarthen*: the animals it invokes were real ones. They belonged to the fauna of the time and would have been known to the poet Aneirin, who wrote the manuscript. So what does *llewyn* mean? Until the most recent bone was discovered in Kinsey Cave (which happens to lie within the region in which Cumbric was spoken), linguists assumed that the word could not have meant what it appeared to mean, so they translated it as wildcat or fox. But the new findings have prompted them to reassess it; it could, after all, mean lynx.[71] (The modern Welsh word for lion, by the way, is *llew*.)

A ninth-century stone cross from the isle of Eigg shows, alongside the deer, boar and aurochs pursued by a mounted hunter, a speckled cat with tasselled ears. Sadly the animal's backside no longer exists: if it had a stubby tail, that might have clinched it.[72] This could be the last known glimpse of the native lynx in British culture. It might have clung on in forest remnants – perhaps in the Grampians – for another few hundred years, but it must have been extinct by AD 1500 at the latest. Like the wolf, it sustained itself in small populations scattered across Europe. Like the wolf, it is gradually emerging from these enclaves.

The lynx does not pursue its prey. It is an ambush predator: it hides beside the places and paths used by the animals on which it feeds, and

---

* Translated by Geraint Jones.[70]

springs on them. Where this species exists, it is a specialist roe deer predator.[73] In the Jura Mountains in Switzerland, for example, almost 70 per cent of the animals lynx kill are roe deer, followed by chamois, fox and hare.[74] Where roe deer are scarce, lynx will kill larger species, such as red deer. Because they are forest animals, seldom leaving the safety of the trees, they present little danger to sheep, unless farmers let their animals into the woods.

There is, as far as researchers can discover, no record, or even an anecdote, of lynx preying on people.[75] They are adept at staying out of sight, and often remain unknown to the humans among whom they live. They are likely to perform a favour for landowners: reducing the populations of deer and foxes. And they could also winkle out the invasive sika deer (introduced from east Asia) which bury themselves in young plantations, where they become inaccessible to human hunters.[76]

Again, according to the leading expert on the subject, David Hetherington, the Scottish Highlands, especially *Am Monadh Ruadh* – commonly called the Cairngorms – are likely to be best suited to the first reintroduction. They have plenty of deer and, thanks in part to their gloomy plantations of exotic conifers, plenty of cover. A smaller population, Dr Hetherington suggests, could be established in the Southern Uplands of Scotland, extending into the Kielder Forest in northern England.[77] The Highlands could support around 400 lynx, he says, which should be a genetically viable population; the Southern Uplands could take around fifty. Unless these regions are connected, by means of wildlife corridors and special passes over the roads, the smaller population is unlikely to sustain itself. New woodlands are being planted fast enough in Scotland to make the reconnection of these places feasible.

Not all reintroductions succeed. Dr Hetherington offers this handy tip for avoiding disappointment: 'Don't do what the Italians did in Gran Paradiso. Only released two lynx. Both male.'[78]

8

A Work of Hope

I'm truly sorry Man's dominion
Has broken Nature's social union,
An' justifies that ill opinion,
    Which makes thee startle
At me, thy poor, earth-born companion,
    An' fellow-mortal!
        Robert Burns
        To a Mouse

I woke to the machine-gun rattle of hail on the windscreen. As I raised my seat, Alan's eyes snapped open. We packed away our lunch and Alan drove back onto the road, then up a track towards the top of the estate. As we climbed, the land became bleaker and darker. The frost-scorched heather was almost black: it looked as if it had been consumed by fire.

We stopped where the road overlooked a little glen in which a few trees grew. As Alan explained why trees had persisted around the streams, I noticed a bird soaring up from the far end of the valley. I was turning away, thinking 'buzzard', when the sun touched the broad planks of its wings. As it flapped towards us, I stiffened in my seat.

'Look!'

The great shoulders, the heavy head, the stout body dispelled my remaining doubts. As it crossed the moor, another eagle plunged down from the sky and dive-bombed it. They rolled over together in the air, then parted and flew on parallel tracks over our heads: two golden eagles, in April. There was, Alan said, a good chance that they were establishing a territory here; perhaps they were already nesting. It was the first time that he had seen a pair on the estate.

We continued up the stony track until we found ourselves among the last lenses of snow filling declivities in the blasted, treeless moor. We left the car. It was bitterly cold. I had made a mistake in assuming that April in the Highlands of Scotland would resemble April in the high lands of Wales. The wind raked through my inadequate clothes. I felt almost naked.

We walked up onto a ridge where tiny twigs of dwarf birch, no higher than my knee, still struggled against the deer. We crawled around in the heather with the wind at our backs, identifying it among the myrtle it resembled. Dundreggan has the greatest concentration remaining in Scotland, but by comparison to the dense dwarf birch tundra I had seen in the Norwegian Arctic, this was unimpressive. The moor was hard and bristly, like an upturned yard brush.

Beside the ridge, Trees for Life had built a large exclosure in 2002, by agreement with the previous owner, to see how the land responded where the deer were excluded. As soon as we stepped into it, I could feel the difference. It felt like walking on a winter duvet: the flora here was soft and spongy. Already a thick sward of pale reindeer lichen, sphagnum and deep grass had formed. The dead stems of bog asphodel still clutched their seed cases.

The land inside the fence was littered with survey poles and transect marks. The scientists Alan worked with had already made discoveries that overturned accepted wisdom. Ecologists had assumed that dwarf birch grows best on boggy land. But here, in the absence of overpopulated deer, the researchers found that it did better on the rocky ridges: other surveys had found more of it on boggy land only because the deer were more reluctant to venture there. Similarly, scientists assumed that the aspen which grows further down the glens prefers steep slopes. But its distribution also appears to be an artefact of overgrazing: as soon as the trees were given some protection, the researchers at Dundreggan discovered that they grew more vigorously on level ground.

Rewilding experiments are likely to present stiff challenges to current scientific knowledge. Many of the places ecologists have studied have been radically altered by human intervention, and many of the processes they have recorded, and which they assumed were natural, appear to have been shaped as much by people and their domestic stock as by wild animals and plants. Like the belief that natural systems are always controlled from the bottom up, now shaken by the discovery of widespread trophic cascades, a number of hypotheses, great and small, could turn out to be false as food webs are allowed to recover.

Alan pointed me to another curiosity. Pushing through the moss and lichen in the exclosure were pine seedlings. Where did they come from? The textbooks, he told me, assert that pine seed tends to travel about fifty metres from its parent tree. But this, he argued, cannot be true of all the seed. At the end of the last Ice Age, pines recolonized Britain from the south. If it takes twenty years for a tree to produce cones, which then spread its seed fifty metres north, Scots pine would not yet have reached London. Yet within 500 years of its return to England, it had arrived in the Lake District. The seed-bearing trees closest to the exclosure were a mile away, and none of the forest creatures that might have carried the cones lived here. Pine must have a means of dispersal that ecologists had so far missed.

It was hard, at first sight, to imagine how it could travel such distances: pine seeds are heavy and their wings are slight. Alan pointed out that when, in the spring, the pine cones crack open, the Highlands are often covered in snow, whose surface melts and then freezes. They are also racked by gales, as I was painfully aware. The shape and smoothness of the seeds suggest, he said, that they might have adapted to ski over frozen snow. I noticed that the saplings in the exclosure mostly grew from crannies or from under large rocks, places in which the seeds might have wedged after skidding over smoother land.

As if to reinforce this idea, the wind howling over the moor suddenly armed itself with frozen snow. Even when I turned my back to the wind I felt as if it were passing straight through me. Then the blizzard stopped just as suddenly and a rainbow arced over the moor. It flashed off again, and just as abruptly we were hit by a squall of rain and hail. Alan, oblivious, had found a heap of black grouse droppings and, stooping over them, had started explaining the ecology of the species. Fascinating as I am sure it was, I decided that I had had enough weather for one day.

As we drove past the little glen, we saw one of the eagles again, planing across the wind. Alan said this was a good sign: if it was holding the territory it was likely to breed here. One predator, perhaps, was already returning.

Here is a table of the large mammals and birds which could be considered for reintroduction into my own country (and which in a few cases have already begun to establish themselves here). Some of the entries might surprise you: I strongly recommend the return of the moose, for example, but not of the wild horse. The wolverine ranks higher on my list than the bear. I have given the grey whale the same score as the eagle owl.

Name of species	Approximate date of extinction in Britain	Suitability for reintroduction	Reintroduction efforts so far
Beaver	No later than the mid-eighteenth century.[1]	10	Officially released into the Knapdale Forest, Argyll. Unofficially released and thriving in the catchment of the River Tay.
Wild boar	The last truly wild boar on record were those killed on the orders of Henry III in the Forest of Dean, in AD 1260.[2]	10	Four small populations in southern England, established after escapes and releases from farms and collections. Likely to spread into other regions if not exterminated.
Elk or Moose (*Alces alces*)	The youngest bones are 3,900 years old, from south-west Scotland.[3]	10	Suitable for reintroduction to forested regions. Released in 2008 into a 450-acre enclosure on the Alladale Estate, Sutherland, as part of a wider rewilding project.

Name of species	Approximate date of extinction in Britain	Suitability for reintroduction	Reintroduction efforts so far
Reindeer	The most recent fossil evidence, from Sutherland, is 8,300 years old.[4]	2	A free-ranging herd grazes on and around Cairn Gorm in the Scottish Highlands.[5] The reindeer belonged to the glacial fauna of Britain and is likely to have become extinct for climatic reasons.
Wild horse	Of the two most recent dates for wild horse fossils in Britain, one has been misreported and the other appears unsafe. The latest now stands at around 9,300 years old.[6]	3	Animals belonging to the last surviving subspecies of wild horse, Przewalski's (*Equus ferus przewalskii*), graze Eelmoor Marsh in Hampshire.[7] Various hardy domesticated breeds are used by conservationists. The question of whether horses should be considered part of our native fauna is controversial, but the evidence suggests that they became extinct largely as a result of climate change.

Name of species	Approximate date of extinction in Britain	Suitability for reintroduction	Reintroduction efforts so far
Forest bison, or wisent	Perhaps soon before the peak of glaciation, between 15,000 and 25,000 years ago.	7	The first herd was released at Alladale in 2011. It has been reintroduced, so far successfully, to a wide range of habitats and climatic zones in Europe and Russia. There are no obvious biological obstacles to its reintroduction.
Saiga antelope	The most recent record is from 12,100 years ago, at Soldier's Hole in Somerset.[8]	1	None. The saiga is an animal of cold dry grasslands, of the kind that existed in Britain towards the end of the Ice Age. It is probably not well adapted to the present climate here.
Lynx	The last known fossil remains date from the sixth century AD, but possible cultural records extend into the ninth century.[9]	9	None. Will take the occasional sheep, so wide-ranging consultation is needed.
Wolf	The last clear record is 1621.[10]	7	None. Should not be introduced without widespread public consent because of a slight risk to people and a higher risk to livestock.

Name of species	Approximate date of extinction in Britain	Suitability for reintroduction	Reintroduction efforts so far
Bear	Not clear. Oliver Rackham and Derek Yalden both suggest around 2,000 years ago.[11]	3	None. Unlikely to be considered seriously unless public safety issues and other conflicts can be resolved.
Wolverine	Derek Yalden suggests 8,000 years ago.[12]	4	Not yet considered. But well adapted for northern and upland regions of Britain. Yalden suggests that, unlike the horse and reindeer, it died out as a result of hunting, not climate change.[13] Likely to kill a lot of sheep. It needs plenty of land.
Lion	The last record of a lion in the region is a bone from an animal that lived in the Netherlands – then still connected to Britain – 10,700 years ago.[14]	1	The cave lion (*Panthera leo spelaea*) was a larger subspecies of the planet's one remaining lion. The clamour for the lion's reintroduction to Britain has, so far, been muted.
Spotted hyena	It died out in Europe around 11,000 years ago.[15]	1	Also likely to face certain political difficulties.

Name of species	Approximate date of extinction in Britain	Suitability for reintroduction	Reintroduction efforts so far
Elephant	The straight-tusked elephant was driven out of Britain by the last glaciation, around 115,000 years ago. It was hunted to extinction elsewhere in Europe around 40,000 years ago.[16] (Another species of elephant, the woolly mammoth, was present in Britain until 12,000 years ago, but that had an entirely different ecology.)	2	The straight-tusked elephant was closely related to the Asian elephant, which might be a good proxy. I have seen no discussion about the reintroduction of elephants to Europe, though I would like to start one.
Black rhinoceros	It never lived here, but two similar species did, the last of which became extinct around 115,000 years ago. The woolly rhino lived here until around 22,000 years ago, and in Germany until 12,500 years ago.[17]	2	The Merck's and narrow-nosed rhinos, which lived in Britain, appear to have been browsing species: they would have eaten trees and shrubs as well as grass. That would make the black rhinoceros a more suitable proxy than the white rhinoceros (whose feeding habit more closely resembles the woolly rhino's, which grazed on grassy steppes during the Ice Age).

Name of species	Approximate date of extinction in Britain	Suitability for reintroduction	Reintroduction efforts so far
Hippo-potamus	Like the elephant, it was driven out by the last glaciation, a little over 100,000 years ago, and later hunted to extinction elsewhere in Europe.	1	None. Our hippopotamus was the same species as the one now surviving in Africa. Suitable habitat in Britain is now in short supply. It can be extremely dangerous.
Grey whale	The most recent palaeontological remains, from Devon, belonged to a whale that died around AD 1610.[18]	7	It appears to have lived in all the seas around Britain before it was hunted to extinction. In 2005, Dr Andrew Ramsey and Dr Owen Nevin of the University of Central Lancashire announced that they planned to fly fifty grey whales from the Pacific to the Irish Sea. 'Some people will say it's impossible, but we are deadly serious about this,' Dr Nevin said.[19] Nothing has been heard of the idea since.
Walrus	Late Bronze Age remains found in the Shetland islands.[20]	2	Walrus are unlikely to have bred in Britain, but appear to have followed their prey here.

Name of species	Approximate date of extinction in Britain	Suitability for reintroduction	Reintroduction efforts so far
European sturgeon	It is not clear when it last bred in British rivers, but this might have been as recently as the nineteenth century. Now critically endangered everywhere, due to overfishing, pollution, dams and weirs.	8	None. Restoring this monstrous fish to its native waters, while difficult,[21] would be a magnificent achievement, and a clear sign that the ecosystems of both rivers and seas were being allowed to recover. There are already reintroduction schemes in the Baltic and North Seas,[22] and attempts to boost the last breeding population in the Gironde-Garonne-Dordogne basin in France.[23]
Blue stag beetle	Probably nineteenth century, as a consequence of deforestation, intense woodland management and the ensuing lack of dead wood.	10	This is a very large and striking metallic beetle. Its reintroduction would depend on ceasing to manage some conservation woodlands. Like many species, it is excluded by systematic coppicing and other measures that break the forest canopy.

Name of species	Approximate date of extinction in Britain	Suitability for reintroduction	Reintroduction efforts so far
White-tailed sea eagle	1916, as a result of persecution and egg collection.[24] Once widely distributed.	10	First introduced to Rum in 1975. They have slowly begun to establish themselves in the islands and west coast of Scotland, and are now being introduced to the east coast of Scotland. An attempt to do the same in East Anglia foundered on opposition from landowners and was stopped by funding cuts.[25] They take a few lambs.
Osprey	1916, partly as a result of egg collectors.[26]	10	Re-established itself in Scotland in 1954, and in Wales in 2004. Introduced to England in 1996.
Eagle owl	The last certain record is from the Mesolithic, 9,000–10,000 years old.[27] But a possible Iron Age bone has been found at Meare in Somerset.[28]	7	Now breeding in some places, after escaping from collections. There is controversy about its impact on other birds of prey, and on pets. Yalden and Albarella in *The History of British Birds* note that 'most birders regard this as a dangerous introduction of a non-native species that should be discouraged . . . All the evidence is to the contrary.'[29]

Name of species	Approximate date of extinction in Britain	Suitability for reintroduction	Reintroduction efforts so far
Goshawk	Wiped out in the nineteenth century, mostly by gamekeepers.	10	Reintroduced in the twentieth century, through a combination of deliberate releases and escapes from falconers. Now around 410 breeding birds in Britain.[30] Still being illegally persecuted.
Capercaillie	1785.[31] The last pair is said to have been shot for a royal wedding banquet at Balmoral.[32]	10	Reintroduced from 1837 onwards. About 2,000 birds remain in Scotland, but they are once more declining rapidly, as a result of cold springs, wet summers and collisions with deer fencing.
Hazel grouse	Though it is likely to have lived here,[33] there is no fossil or cultural evidence beyond the late Ice Age.	3	None.

Name of species	Approximate date of extinction in Britain	Suitability for reintroduction	Reintroduction efforts so far
Great bustard	Last known breeding pair: 1832 in Suffolk.[34] Hunted to extinction.	9	Reintroduced in 2004 to Salisbury Plain. Slowly spreading into other parts of England. Though it is a steppe species, and might have become extinct here even without the help of humans as a result of climate change, it seems happy to live in arable land, so there is plenty of potential habitat.
Common crane	Last evidence of breeding in Britain: 1542.[35] The many Cran- place names in Britain indicate their presence.	10	Cranes re-established themselves through migration in the Norfolk Broads in 1979, and have bred there since then. Now breeding in two other places in eastern England. Re-introduced in 2010 to the Somerset Levels.[36]

Name of species	Approximate date of extinction in Britain	Suitability for reintroduction	Reintroduction efforts so far
White stork	Last recorded nesting in Edinburgh in 1416.[37]	10	In 2004 in Yorkshire a pair tried to breed on an electricity pole, whose cables happened to have been turned off for maintenance.[38] In 2012 a lone bird built a nest on top of a restaurant in Nottinghamshire.[39] There are frequent visits by non-breeding birds.
Spoonbill	The last breeding records until recently were 1602 in Pembrokeshire and 1650 in East Anglia.[40]	10	A breeding colony of six pairs established itself at Holkham in Norfolk in 2010. In 2011, the colony rose to eight pairs and produced fourteen young.[41] In 2012 nine pairs bred, and nineteen young birds fledged.[42]
Night heron	Last bred here in either the sixteenth or seventeenth century, at Greenwich. Believed to be the *brewes* or *brues* often served at medieval banquets.[43]	10	Today it is a scarce visitor. It currently breeds in many parts of Europe. Now that it is no longer persecuted in Britain, it may start breeding again here.

Name of species	Approximate date of extinction in Britain	Suitability for reintroduction	Reintroduction efforts so far
Dalmatian pelican	Remains have been found from the Bronze Age in the Cambridgeshire Fens and from the Iron Age in the Somerset Levels, close to Glastonbury. A single medieval bone has been found in the same place.[44]	10	None. The pelican's range, which once covered much of Europe, has steadily shrunk. It is sensitive to disturbance, and its habitat has been reduced by drainage. Two thousand years ago, Pliny recorded that it was still breeding on the Rhine, Scheldt and Elbe rivers.[45] Today the nearest breeding colonies are on the Danube and in Montenegro. This means that pelicans are unlikely to recolonize Britain naturally: they would have to be introduced.

This list is offered as a catalogue of plausibility. The highest scores represent the reintroductions that might be tried first, on the grounds that they are most likely to succeed, to be politically acceptable and to help restore dynamic processes in the rewilding lands or seas of this country in the current (and warming) climate. Polar bears need not apply.

Once such species have been established at genetically viable population sizes and protected from man-made hazards, they should, more or less, be left to get on with it. If they cannot survive here, that answers the question of whether or not the reintroduction was appropriate.

Broadly speaking, I have marked down the Ice Age and Preboreal species – those adapted to the open tundra or steppes, the habitats available during and soon after the great freeze. If an animal died out as a result of warming and the habitat changes this caused, it is likely to be less suited to the current climate than those which may have been hunted to extinction. This is why I have judged the reindeer and horse harshly: they returned to Britain soon after the glaciers retreated, but disappeared as the grasslands of the cold, dry Preboreal period that followed gave way to forest.

We cannot always be sure which factor was most important in the disappearance of an ancient species. Some of them would have been affected by both climate change and hunting. So we must make educated guesses, comparing the survival of the horse and the reindeer, for example, to that of other hunted species, such as the moose, the aurochs and the red deer, which lasted much longer. The question of whether horses and reindeer disappeared because the grasslands turned to forest or the grasslands turned to forest because horses and reindeer disappeared is also hard to resolve. But even those who conducted the research proposing that the northern Siberian steppes turned to tundra because the grazing animals were killed by hunters suggest that the southern steppes turned to forest for climatic reasons.* Nor do we have definitive extinction dates, as the fossil record is far from complete.

My aim here is to expand the range of what we consider possible, to open up the ecological imagination. That requires some understanding of palaeoecology. The fact that sometimes eludes biologists and naturalists, steeped in the present, is that every continent except Antarctica possessed a megafauna.

When I studied zoology at university, I read a number of accounts, founded on ecology and physiology, which tried to explain why very large animals live in the tropics but not in temperate nations. I found them interesting and in some cases persuasive. But, like the authors of these speculations, I had missed something. The inherent difference they sought to explain did not exist. Until very recently, large animals lived almost everywhere, often in great numbers. They could do so

---

* Zimov et al maintain that 'boreal forest expanded northward at the end of the Pleistocene into areas that had been predominantly steppe, presumably in response to climatic warming'.[46] Elsewhere Zimov writes: 'In the southern steppes, the situation is different. There, the warmer soil allows for more rapid decomposition of plant litter even in the absence of herbivores.'[47]

today: African lions have been living and breeding in outdoor enclosures in Novosibirsk zoo in Siberia since the 1950s. Large animals appear, in most parts of the world, to have been hunted to extinction by people. These species have been excluded from temperate regions not by any natural ecological or physiological constraints, but by humans.

With the possible exceptions of Australia's and Madagascar's, none of these megafaunas has the capacity to amaze as much as that of the Americas. Alongside mammoths of several species (including one that dwarfed the woolly variety), mastodons, four-tusked and spiral-tusked elephants, lived an improbable bestiary of other massive herbivores. There was a beaver (*Castoroides ohioensis*) the size of a black bear: eight feet from nose to tail, with six-inch teeth. There was a giant bison (*Bison latifrons*) whose bulls weighed two tonnes, stood eight feet at the shoulder and carried horns seven feet across. Shrub oxen (*Euceratherium collinum*) and musk oxen inhabited the entire northern continent. (Neither of them are really oxen: they are closely related to sheep and goats, but very much larger.) In South America there was a giant llama (*Macrauchenia*) whose face ended in a trunk. There were armadillos – glyptodonts, such as *Glyptodon* and *Doedicurus* – the size of small cars, armoured with a bony carapace like a tortoise's. Ground sloths – such as *Megatherium* and *Eremotherium* – the weight of elephants stood twenty feet on their hind legs, and used their formidable claws to pull down trees.

The great American lion (*Panthera leo atrox*), one of the largest cats ever to have existed, was almost sweet by comparison to the terrifying *Smilodon populator* – the giant sabretooth cat – which weighed as much as a brown bear, hunted in packs and possessed fangs a foot long. The short-faced bear (*Arctodus simus*) stood thirteen feet in its hind socks; the Riverbluff Cave in Missouri has scratch marks made by its claws fifteen feet from the floor.[48] One hypothesis maintains that its astonishing size and shocking armoury of teeth and claws are the hallmarks of a specialist scavenger: it specialized in driving giant lions and sabretooth cats off their prey.[49]

The North American roc (*Aiolornis incredibilis*), had a wingspan of sixteen feet and a hooked bill the length of a man's foot. No skull of another predatory bird, the Argentine roc (*Argentavis magnificens*) has yet been found, but the available bones suggest that its wings were twenty-six feet across and that it weighed twelve stone.[50] On the

Pacific coast, sabretooth salmon (*Oncorhynchus rastrosus*) nine feet long migrated up the rivers.

All these remarkable beasts disappeared at around the same time. generally between 15,000 and 10,000 years ago. Their extinction coincides with the arrival and dispersal of the first technologically sophisticated people in the hemisphere: hunters using finely worked stone weapons. The evidence suggests that it was not, as many palae-ontologists first supposed, primarily climate change that wiped out the American megafauna:[51] it had survived massive fluctuations in the recent past, and the habitats that many of the missing species required still exist. They were hunted to extinction.*

The animals of the New World had never encountered humans before, except perhaps some scattered bands with basic technologies. So, like the unfortunate beasts of the islands discovered by Europeans, they probably stood and watched, without fear, as the hunters approached.

Had the Mesolithic people of the Americas eaten everything they killed, they would scarcely have trimmed the herds of game, so small were their numbers. One ground sloth could have fed a clan of hunters for months. The speed with which the megafauna of the Americas collapsed might suggest that they slaughtered everything they encountered.†
Among those who broke into the New World, anyone could be a Theseus or a Hercules: slaying improbable monsters, laying up a stock of epic tales to pass to their descendants. Like all those who have discovered wildlife in its unexploited state – the sailors who found the dodos in Mauritius or the whales in the southern oceans, the fishermen who first assayed the Grand Banks off Newfoundland – they might have thought the sport would last for ever. Perhaps the care with which some indigenous people of the Americas engage with the natural world came later.

---

* William Ripple and Blaire Van Valkenburgh caution that the populations of large herbivores are likely to have been low, as they were suppressed by predators and subject to trophic cascades. This could have made it easy for humans to have driven them to extinction.[52]

† Again, it is worth bearing the alternative hypothesis in mind: that the herbivores could have been tipped into extinction easily, as their numbers were low. If people deprived other predators of their largest prey, those predators would have been forced to kill smaller animals (as wolves in Alaska do when hunters have reduced the moose population). This might have created a powerful knock-on effect, as extinctions cascaded down the food chain.

Slaughter of this kind revolts us, but are not most of our great myths built on such adventures? Do Ulysses, Sinbad, Sigurd, Beowulf, Cú Chulainn, St George, Arjuna, Lâc Long Quân and Glooskap not survive in a thousand current tales? All of us have ancestors who, regardless of the continent they inhabited, must have battled with beasts many times their size, armed with horns and tusks and claws and fangs, and must have passed down tales of their triumphs and tragedies, sagas which mutated and evolved across hundreds of generations, but which maintain their essential form today. Are these struggles with the beasts of prehistory not imprinted in our subconscious as surely as Homer's epics were eventually committed to papyrus?

To re-enact these quests, the Romans scoured Africa for monsters to release into their amphitheatres. The Spanish breed black bulls with the temperament of giant aurochs. The Maasai risk long prison terms, mutilation and death to hunt lions. Societies throughout Europe engaged until recently in cruel sports involving bears, badgers, dogs – any creature fierce enough to reawaken the ancestral thrill. The absence of monsters forces us to sublimate and transliterate, to invent quests and challenges, to seek an escape from ecological boredom.

An interesting question arises. Why, when the megafauna was eliminated in the Americas, in Australia, New Zealand, Madagascar and Europe, does it survive, at least in part, on mainland Africa and in some places in Asia? There creatures exist which, were we not familiar with them, would invoke the wonder and incredulity with which we contemplate the glyptodont, the elephant bird and the marsupial lion. Elephants, rhinoceroses, giraffes, hippos, eland, cheetahs, tigers: all of them, had they lived in other parts of the world, would have been – or were – exterminated. The answer is surely that in Africa and southern Asia, they evolved alongside hominids and early humans. They learnt to fear the insatiable ape, the diminutive monster which could look back upon its deeds and forward to their embellishment.

People who call themselves Pleistocene rewilders seek to recapitulate the prehuman fauna of the Americas.[54] They point out that the extinctions terminated trophic cascades and other processes that must have shaped the ecosystems of the New World. Species which evolved alongside the missing megafauna, such as the pronghorn, whose remarkable speed – up to sixty miles per hour – is likely to have been an adaptation to the presence of the American cheetah, now inhabit an ecological vacuum, in which they are constrained by neither predation nor competition. These rewilders call for the introduction of proxy species to the Americas: exotic members of the groups that became extinct, or animals which fulfil a similar ecological role.

They talk of introducing Bactrian camels, which live in central Asia, to replace a similar animal, *Camelops*, which lived in large numbers in North America until humans arrived. They suggest importing the African cheetah to hunt pronghorns, the African lion to pursue feral horses (which, now widespread, are good proxies for the wild horses which once roamed the continent), African and Asian elephants to replace the mammoths, mastodons and other such monsters. (Perhaps Americans should be grateful that there is no living substitute for the giant sabre-tooth or the short-faced bear.) Not only, they argue, would these beasts help to revive American ecosystems and heighten people's interest in conservation and rewilding, but they would also be better protected from extinction if they were living in the wild on more than one continent.

It would not be correct to report that these proposals have been greeted with universal enthusiasm in North America. Aside from obvious concerns about the release of lions and elephants, some ecologists have objected that superficial similarities can mask major genetic differences: the American cheetah (a larger animal than the African species) was more closely related to the puma, for example.[55] The proxy species evolved in some cases in response to ecosystems and climatic conditions different from those that prevailed in America before humans arrived. It would be surprising if the way in which they engaged with the remnant American ecosystem closely mimicked the ecological relationships of the species they are supposed to replace. But the idea is worthy of investigation, and perhaps a few experiments.

There are fewer biological obstacles to the reintroduction of a missing megafauna to Europe. Unlike the extinct American beasts, the monsters which once ranged across this continent have close relatives in Africa or Asia. The hippos submerged in Trafalgar Square were of the same species, *Hippopotamus amphibius*, that lives in Africa today. It survived in parts of Europe until around 30,000 years ago, when it appears to have been hunted to extinction.[56] The last temperate rhinoceros species to disappear from the continent bear some resemblance to the black rhinoceros, which is likely to fill a similar ecological niche. The Asian elephant might be a good proxy for its relative the straight-tusked elephant.

Reintroducing elephants to Europe would first require a certain amount of public persuasion. To find enough forage, wild elephants would have to make long migrations, especially in the winter. Gardeners, farmers and foresters are unlikely to applaud the proposal, though it would take our minds off the slugs and aphids with which so many of us are obsessed. But if very large areas of land are allowed to rewild as farmers depart, it would be a pity not to remember and at least consider the most powerful of our missing species.

The Pleistocene Park being established in north-eastern Siberia by Sergey Zimov and other visionary ecologists is, most of the time, less contentious. The rewilders began, in 1988, by releasing Yakutian horses – believed to be closely related to the wild horses that lived in the region towards the end of the Ice Age – into a park of 160 square kilometres (the size of Liechtenstein). Reindeer, moose and wild snow sheep (similar to the North American bighorn) already lived in the area, as well as lynx, wolves, bears and wolverines. Since then, musk oxen, forest bison and red deer have been reintroduced.[57] At some point the park will be expanded by a further 600 square kilometres, becoming a little larger than the island of Minorca.

Zimov and his team are either considering or being urged to consider the introduction of several other species which once lived in the region or which are closely related to those that did. Among them are saiga antelope, Bactrian camels, Amur leopards, Siberian tigers and lions. Already, as Zimov's experiments predicted, the new grazers are turning the moss and lichen tundra into grassy steppe. The question of whether this transition will accelerate climate change needs to be carefully examined. His assumption that the restoration of grassland will reduce global warming could be optimistic,[58] and has been partly contradicted by no less an authority than, er, Sergey Zimov,* lead author of a paper written ten years earlier.[60]

---

* Zimov and colleagues now argue that because the steppes are drier than mossy tundra, they are less likely to generate and release methane, a powerful greenhouse gas. Being paler, they also absorb less heat.[59] But these effects will be at least in part counteracted by the effect he documented in 1995: moss insulates the soil much more effectively

Some people appear to be giving serious consideration to the idea of restoring another missing member of the Siberian ecosystem. Whatever the drawbacks may be, the notion (which might or might not be fanciful[61]) of resurrecting the woolly mammoth by extracting genetic material from frozen corpses and injecting it into the eggs of Asian elephants possesses the virtue of firing the imagination on all cylinders. But it seems odd that, while there has been so much attention and money given to this project, the idea of simply reintroducing the Asian elephant to parts of Europe and Asia, from which it or its sibling species (the straight-tusked elephant) has been extirpated, has not yet taken root; or even, as far as I can discover, been discussed. The elephant in the forest – the huge and obvious fact that almost everyone has overlooked – is the most prodigious instance of Shifting Baseline Syndrome I have chanced upon so far. Who knows what else we might all have missed?

The North American debate raises another important question, which is relevant everywhere: is a healthy and desirable ecosystem necessarily composed of native species? Certain exotic animals and plants destroy ecological diversity of all kinds in the places they infest. Without natural predators or parasites or diseases, attacking native species which have evolved no defences against them, they can quickly overwhelm an ecosystem, sometimes to the point at which (as I have seen in small streams in England infested by American signal crayfish) the last robust ecological process still taking place consists of big ones eating little ones.

In some places the progress of these invasive species looks like the plot of a Gothic novel. The walking catfish, for example, native to south-east Asia, has escaped from fish farms and ornamental ponds in China and the United States, and now crawls overland at night, colonizing water that no other fish can reach.[62] It eats almost anything that moves. It slips into fish farms and quietly works through the stock. It burrows into the mud when times are hard and lies without food for months, before exploding back into the ecosystem when conditions improve.

The cane toad, once confined to Central and South America, has been widely introduced in the tropics to control crop pests. Unfortunately it also controls many species which are not considered pests. It

---

than grass, preventing the permafrost from thawing and releasing the methane and carbon dioxide it contains. It is not clear at this stage which effect will dominate.

appears to be almost indestructible: one specimen was seen happily consuming a lit cigarette butt.[63] Scarcely anything which tries to eat it survives: it is as dangerous to predators as it is to prey. Unlike other amphibians, it can breed in salty water: it could have waddled out of the pages of Karel Capek's novel *War with the Newts*.

The world's most important seabird colony – Gough Island in the South Atlantic – is now being threatened by an unlikely predator: the common house mouse. After escaping from whaling boats 150 years ago, it quickly evolved to triple in size, and switched from eating plants to eating flesh. The seabirds there have no defences against predation, so the mouse simply walks into their nests and starts eating the chicks alive. Among their prey are albatross fledglings, which weigh some 300 times as much as they do. A biologist who has witnessed this carnage observed that 'it is like a tabby cat attacking a hippopotamus'.[64]

But even more mundane invasions can be devastating to the richness of native ecosystems. *Rhododendron ponticum*, which – as the name suggests – is native to the shores of the Black Sea and lands at similar latitudes, works its way through British woodlands, smothering and poisoning other plants. It can kill even the mature trees among which it grows. I have seen entire stands of ash dying from canker, apparently as a result of the moist conditions sustained around their boles by the rhododendron's thick cover. It harbours sudden oak death fungus, which kills a number of trees in Britain, though not, as it happens, oaks. While the hawthorn in Britain supports 149 species of insect, the birch 229 and the oak 284, the rhododendron is reported to harbour none.[65]

This is one of the reasons why it thrives here: it has escaped from the restraints imposed by the plant eaters of its native lands. Interestingly, however, *Rhododendron ponticum* was native to these islands during a previous interglacial period.[66] Its natural pests, predators and competitors appear to have been destroyed by subsequent ice advances, allowing it, once imported by enthusiasts, to return here unchallenged, our flora's *deus invictus*. Is it possible that one of our missing herbivores – the ancient elephant or the Merck's or narrow-nosed rhino, for example – was able to eat it? If it is not controlled, it will eventually supplant almost all the vegetation of the places it invades.

I am struck by how unassuming some of the species which cause havoc abroad are in their native range. In the Himalayas where it belongs, and where despairing householders might fervently wish it had stayed, dry rot is a fungus living on pine and yew trees. It is so rare that between 1953 and 1992 it was officially recorded only three times,[67] and it may be in danger of extinction in the wild. In Britain, purple loose-strife is an occasional and delightful native ornament of our riverbanks and lakesides. In North America and New Zealand it is a rampaging, uncontrollable menace, smothering wetlands and choking rivers.

But there are many exotic species which cause little discernible harm to the countries they colonize. Until recently I had not realized that the little owl does not belong to our native fauna: it was intro-duced to Britain in the nineteenth century. But its presence here is uncontroversial: it persists in fairly small numbers without driving out native species. The knowledge that it did not originate here will do nothing to dampen my delight next time I see one.

Many of the plant species – 157 according to one estimate[68] – that we once saw as native now appear to be what botanists call archaeophytes: exotic species which arrived before the year 1500. A handful reached Britain during the Neolithic, their seeds probably lurking in the grain brought here for sowing by the first farmers, or, perhaps, stuck to the feet of travellers or in the hides and fleeces of the animals they imported.

Some archaeophytes are familiar to anyone who loves nature, and their inclusion on the list of non-native species is often surprising: field poppy, greater burdock, cornflower, wormwood, scarlet pimper-nel, shepherd's purse, fumitory, corncockle, deadnettle, common mallow, crack willow, common vetch, field pansy, mayweed and white campion, for example.[69] You can find several of them in the packets of wildflower seeds we are encouraged to sow in spare corners of our gardens, to save Britain's native flora. As their lovely names suggest, they have seeded themselves in our culture and are as embedded in our lives as the species that arrived before we did.

Among these archaeophytes are plants which are now extremely rare. The pheasant's eye, for example, which appears to have arrived in the Iron Age, is marked as endangered on Britain's Red Data List,[70] and officially classed as a priority species for conservation here.[71] Is it illogical to seek to save these plants, even in the knowledge that they were brought here by humans? They do no harm and afford delight and wonder to those who appreciate them, which is surely all that is required to make something worthy of preservation. Even so, it compounds the confusion – seldom acknowledged, let alone resolved – between conservation and gardening.

Some animal species might also have been mistakenly seen as native. The eminent mammal biologist Derek Yalden presents compelling evidence that the brown hare was brought here by people.[72] The bones of what appear to have been mountain hares (a different species) are found in England and Wales in deposits from the early Mesolithic, soon after the ice sheets retreated. They appear to have been driven out (perhaps surviving in Scotland) as the land became forested. Possible records of brown hares begin to appear in the Bronze Age; more certain remains in the Iron Age. In *Commentarii de Bello Gallico*, Julius Caesar records that the Britons considered hares, fowl and geese 'unlawful to eat, but rear them for pleasure and amusement'.[73] This raises the possibility that brown hares were brought to Britain either as pets or to be hunted for sport.

Restoring a functioning ecosystem does not equate to purging all non-native species. It requires only that we control or suppress those species which deprive many others of a foothold here. Even some of the most prolific exotic animals could be subdued by native predators. Grey squirrels, for example, are currently storming through the ecosystem, defying attempts by humans to restrain them. Ecologists hate them, with good reason. But pine martens and goshawks love them[74] (in the purely carnal sense). Had landowners not waged war on all predators, regardless of their impacts, they might not have had to wage – and lose – the current war against grey squirrels. Martens and goshawks, now returning to some of the places from which they were exterminated, may have the potential to reduce the grey squirrel to such an extent that it begins to function ecologically much as a native species would.

Where rivers contain healthy populations of predatory fish, they appear to thrive on invasive crayfish. Sometimes when I have caught a fat perch for my dinner, I have found a crayfish or two in its stomach. Perhaps because of the acidity of the fish's stomach, the shell dissolves before the flesh does: I have extracted from the insides of a perch perfectly peeled crayfish tails, which look as if they have just been shovelled off a fishmonger's slab. I have noticed that where large chub lurk, the crustaceans are more reluctant to emerge from under their stones. Twice in my crayfish nets I have trapped enormous pike – one of which must have weighed well over twenty pounds and pulled the net all over the river as I tried to retrieve it – though I cannot say whether they had come after the crayfish or the bait. I would be surprised if these aquatic locusts were not also consumed by barbel, trout and eels. It is possible that in places where pollution levels are low enough for fish to thrive, these predators will eventually suppress the crayfish population until it ceases to threaten some of the native wildlife it now displaces.

Both otters and polecats, native to Europe, appear to drive American mink out of their territories.[75] In the Finnish archipelago the white-tailed sea eagle, now recovering from near-extinction, also seems to be reducing the mink's range.[76] This great eagle, recently reintroduced, could have the same effect in Britain.

Even so, invasive species challenge attempts to defend a unique and distinctive fauna and flora. Certain animals and plants have characteristics that allow them to invade and colonize many parts of the world, and there is a danger that ecosystems everywhere come to contain a similar set of species, making the world a blander and less surprising place. Even if they are suppressed by predators, grey squirrels and red signal crayfish will continue to destroy their competitors (red squirrels and white-clawed crayfish) by exposing them to the diseases they carry. We should try to prevent them from spreading further, but accept that they cannot be eradicated: grey squirrels, mink and signal crayfish now belong to ecosystems from which they used to be absent, and the best we can hope for is that they are firmly sat upon by other species.

On the day after our foray into Dundreggan, Alan took me to Glen Affric, which is said to contain the least altered large area of woodland in Britain.[77] It was a bitter, wet day. From the road along the valley of the river Affric the old forest looked like a giant tray of broccoli. When Scots pine is young, it is slim and pointed. But the mature trees spread out into a broad, rounded canopy. The road wound round bluffs to which the ancient trees clung, their crabbed and twisted shapes reflected in the fissured rocks.

We stopped above a waterfall whose cool breath I could feel while standing on the rocks over the gorge, and whose spray I could taste on the air: mossy, halogenic. The peaty brown water stretched dark olive over the sill before plunging and pluming down the long series of rapids. The gorge was a Japanese painting, knotty pines bristling on crooked rocks above the water.

On the far bank, preserved from grazing, the boulders beneath the trees were carpeted in moss and lichen, through which cowberry and bilberry grew. Around them the heather sprawled in deep drifts. The trees too, in the perpetual mist raised by the falls, were bearded and maned with outrageous growths of lichen. The hazels and rowans in the understorey scarcely emerged from their shawls of moss. This, Alan reminded me, was rainforest.

The road took us past Loch Beinn a Mheadhoin, whose waters looked like brushed steel. On its islands and bluffs grew umbrella-shaped pines. Beneath them, inaccessible to the deer, young trees spiked towards the light.

Glen Affric is one of the few parts of Britain in which the work of the Forestry Commission has, from the beginning, been largely benign. Since a sawmill was built in the valley in 1750, the old trees had been under siege, while the sheep grazing beneath them prevented almost all recruitment. The commission bought most of the glen in 1951, and, neglecting its customary duties, decided to preserve it rather than to wreck it. In the 1960s a young forester persuaded his bosses to let him fence 800 hectares of the glen, arguing, against the received wisdom of the time, that the trees could regenerate without being planted.

The results were spectacular, an unequivocal rejoinder to those who said it was impossible. We could see them on the brae on the far side of the loch: stockades of pines a few decades old, their spiky profile broken in some places by the great humps of older trees. This experiment was one of the factors that had inspired Alan to found Trees for Life.

He parked the car at the head of the loch, in a patch of birch and pine wood. Here, by contrast to the fissured grey bark in Glenmoriston, the trunks of the birch trees were mostly white and smooth. Beneath them he pointed out something that fascinated me. The ground was covered in hummocks, which I might have taken for anthills. Alan explained that they were growths covering rocks and old tree stumps. Springing between the humps, he showed me the successional process. After a rock rolls down from the slopes above or is bared by disturbance, lichens begin to creep over it. They dissolve some of the mineral content, breaking down the surface and creating organic matter. This allows moss to move in, displacing the pioneer lichens. The moss in turn creates a habitat for leafy plants such as bilberry and cowberry. The process can take a century or more. These hummocks are a characteristic feature of old forest. They will form only under trees, perhaps because in such thin soil the plants would dry out in the open. Alan had watched one rock for twenty years and seen the vegetation it harboured shifting from one phase to the next.

After he had pledged to restore the Caledonian Forest in 1986, he spent a couple of years educating himself and raising money. He began by persuading some private landowners in Glen Cannich, to the north of where we stood, to allow him to protect pine seedlings on their estates. In 1989 he took a Forestry Commission official to a place in Glen Affric in which remnant pines were growing.

'I said, "You've got the land, we've got the money. Let's put them together." It was an unlikely partnership. I was a hippy-like character from Findhorn with a beard and long hair, he was a government official. But the relationship between Trees for Life and the commission has been going strong ever since.

'We're more radical than they are. They can't take a position on wolves, for example. Nor are they ready to embrace the removal of roads and tracks – yet. We can be bolder than them. I know the glen better than many of their staff, and I can see opportunities which sometimes they haven't yet spotted. About three-quarters of the trees we've planted are on Forestry Commission land, on many of its estates across the Highlands. We're working with their neighbours as well. The idea is to connect the new forests all the way to the west coast.'

We set off along the track on foot, then soon plunged into deep heather and struck up the hillside. The great pines here, none younger than a century, looked like the acacias of East Africa, flat-topped above the dun savannahs. Some were wider than they were tall. Each had a distinct growth pattern. Some trees had a single straight trunk, unbranched until it spread into the canopy; some had branches all the

way up; some possessed multiple trunks; one or two grew almost horizontally. Their trunks were elephant grey, their branches dragon-scaled
in sunset pink, crowned with a haze of shrubby needles.

'I call it the geriatric forest. It's like an old people's home. The deer
come down here in the winter. As soon as the seedlings reach the
height of the heather, they get eaten.

'The problem is not deer. It's the stalking industry, which ensures
that the deer are overpopulated. The Forestry Commission has sporting tenants. They don't live here, they just come to shoot the deer, but
they hamstring us. Their attitudes are very traditional. One of them,
an Englishman, threatened to burn my house down.

'Red deer in Scotland are about two-thirds of the size of those in
continental Europe, and of those preserved in peat bogs here. They
are woodland creatures. On open ground they have less to eat. The
deer in the Highlands are the runts of the glen. When settlers in North
America saw the red deer there, they were so much bigger than the
British specimens that they assumed they were a different species and
called them elk. It's been a source of confusion ever since.' (It now
appears that, though very closely related, they *are* a different species:
the North American red deer (or elk) was reclassified in 2004 as *Cervus canadensis*. Another possible reason for the reduction in size is
that hunters tend to select and kill the biggest stags.)

I later read that *The Monarch of the Glen*, painted by Sir Edwin
Landseer, who also sculpted the recrudescent lions in Trafalgar Square,
was set in Glen Affric. (The location is hotly disputed, however. Other
accounts suggest that it was painted in Glenfeshie, Glen Orchy or Glen
Quoich.) Completed in 1851, the painting became the emblem of the
ersatz culture, the Balmorality, created in the newly cleared Highlands
by Victoria and Albert at Balmoral Castle and by the aristocrats who
mimicked them. This mythologized re-enactment of the lives of the vanished Highland peoples – all tartans and claymores – was the narrative
with which those who had expropriated the land and expelled its inhabitants justified and eulogized the new dispensation. It was the Scottish
equivalent of Marie-Antoinette's Hameau de la Reine, at Versailles.

The painting depicts a magnificent stag, overfed and splendidly
pointed, eyes raised imperiously to the hills: both the idealized quarry

of the new lairds and their own imagined embodiment. It stands on a mountaintop surrounded by bare hills. The pose, gaze and setting bear, to my eyes, a striking similarity to Franz Winterhalter's 1842 portrait of Prince Albert. There could scarcely be a greater contrast with either the squalid reality of dispossession and seizure or the weedy, stunted deer living there today.

As the freezing rain worked its way through my thin coat and worn-out boots, we came to a high fence, and passed through a gate which seemed like a door to another world, so great was the contrast between the vegetation on either side of it. This was the fence which, in 1990, the Forestry Commission agreed to erect around fifty hectares of brae, using the money that Alan and the Findhorn Foundation had raised. On one side the grass was nibbled low and covered in deer droppings. Apart from a few small saplings buried in the heather, and one or two growing out of reach of the deer in the crooks of fallen trunks, there were no young trees. On the other side was a mosaic of habitats of the kind that, Alan said, we could expect to see regenerating across the Highlands if deer numbers were reduced.

The wet ground was thick with bog myrtle, which in the summer would fill the air with its drowsy scent. Here the pine seedlings had crept up, agonizingly slow. Young conifers are easy to date: each star of branches growing from the trunk denotes one year's growth. These trees, no higher than my chest, some below my waist, turned out, when we counted the layers, to have germinated when the fence was erected. Apart from their size, they looked like the mature trees on the other side of the fence: they had developed, in miniature, the same range of growth patterns.

'They're bonsai trees. The Japanese mimic nature: growing trees in adverse conditions like these.'

But on the drier ridge just a few yards away, the trees had been growing as fast in two years as some of those on the boggy ground had grown in twenty. The highest was now twenty-three feet tall (Alan told me that this specimen had been the focus of his affection, and that you could see the difference this made). They grew straight and sharp; it would be several decades before they began to acquire the hunched and spreading individuality of the bonsai bog trees. Among them were rowans of twice my height and more, and regenerating birch and juniper. An orchid rare outside the exclosure – creeping ladies' tresses – had proliferated here.

The old trees within the fence were now dying quickly. Several had collapsed and would be left where they fell. The resin they contained would prevent the trunk from disappearing for around a century. Others had died in their boots and were now shedding their leafless twigs. The dead trees would provide habitats for species which cannot survive on living wood: fungi, certain lichens, beetles, pine hoverflies, birds – such as owls, woodpeckers and crested tits – and bats, which nest in holes in the rotting wood. As they decay, they release a steady trickle of nutrients which other plants can use.[78]

'I like to think the trees know they can go now, as they've done their bit, and their children are growing up around them.'

Alan told me that they would exclude the deer for a few more years, then they would reduce the height of the fence in some places, and let a few in. 'Deer should be a part of this system, but not in such numbers.' When deer numbers were reduced across the Highlands, the exclosures would be removed.

In places like this, where some living trees had clung on, the rewilders could let nature do the work. In others, like the bare West Affric estate, bought by the National Trust for Scotland partly as a result of campaigning by Trees for Life, they had to plant islands of forest, grown from the nearest seed sources, trying to replicate the patterns and distributions in which trees might have grown there naturally, to begin the process of regeneration. Alan's intention was to re-seed native forests along the glens that struck diagonally across the Highlands, then to connect them through passes low enough to lie beneath the treeline.[79] He described the pine as a crucial species, which creates the habitat required by much of the missing native wildlife. Some would return naturally. Other species – from the wood ant to the wolf – would have to be brought to the forests and released.

Alan already appears to have catalysed a gradual rewilding of the entire watershed of the River Affric. This will, if the plans mature, create a corridor of native forest twenty-five miles long.[80] But this is just one corner of the 1,000 square miles whose ecosystems he seeks to restore.

'One of the things I've learnt,' he told me, 'is patience. We're talking about trees with a lifespan of 250 years or so. That's not so long. In

California, it would take 2,000 years to regrow mature redwoods. And it's easy here compared with other places. In Nepal the soil is washing off the slopes of the Himalayas as a result of deforestation; so much that it's forming an island in the Bay of Bengal. Here the soil is acidified and low in nutrients, but we've still got it, which is why rewilding will take only 250 years.'

Some of the major landowners in the region were hostile to his ideas, seeing them, correctly, as a threat to the universal application of the land use they favoured: intensive grazing by deer or sheep, supported by stalking fees or farm subsidies. But, he says, attitudes on some estates are slowly changing. Attitudes among other Scottish people are changing much faster.

'We've tolerated the absentee landlords with scarcely a murmur of discontent. Scotland suffered a huge psychological blow as a result of the loss of the Battle of Culloden. It is still a psychological wound in the nation today. The Clearances happened partly as a consequence. They brought the sheep in and cleared the people off. Scotland became subservient and demoralized. We became a nation of sheep. Like all indigenous people when they lose their connection to the land, we lost our confidence.

'But over the past twenty or thirty years there has been a tremendous reawakening of our engagement with the land. You can see it in the number of people here who have joined woodland groups or who go hillwalking. Now people know about the Caledonian Forest. It has gone hand in hand with the increased political awareness which led to the creation of the Scottish parliament. It's a small step to recognizing that we need to care for the land. But how can we do so if it doesn't belong to the people who live here?'

As the rain seeped through my coat, down my trouser legs and into my boots, and I found myself wishing that he would show some sign of the discomfort I was feeling and some inclination to walk down the hill and get back in the sodding car, Alan voiced the thoughts that had, over the past few months, been forming in my mind: 'The environmental movement up till now has necessarily been reactive. We have been clear about what we don't like. But we also need to say what we would like. We need to show where hope lies. Ecological restoration is a work of hope.'

9

Sheepwrecked

By Langley bush I roam but the bush hath left its hill
On cowper green I stray tis a desert strange and chill
And spreading lea close oak ere decay had penned its will
To the axe of the spoiler and self interest fell a prey

John Clare
Remembrances

Most human endeavours, unless checked by public dissent, evolve into monocultures. Money seeks out a region's comparative advantage – the field in which it competes most successfully – and promotes it to the exclusion of all else. Every landscape or seascape, if this process is loosed, performs just one function.

This greatly taxes the natural world. An aquifer might contain enough water to allow some farmers to grow alfalfa, but perhaps not all of them. A loch or bay or fjord might have room for wild salmon and a few salmon farms, but if too many cages are built, the parasites which infest them will overwhelm the wild fish. Many farmland birds can survive in a mixed landscape of pasture and arable crops, hedgerows and woodlands, but not in a boundless field of wheat or soya.

Some enthusiasts for rewilding see reserves of self-willed land as an exchange for featureless monocultures elsewhere. I believe that pockets of wild land – small in some places, large in others – should be accessible to everyone: no one should have to travel far to seek refuge from the ordered world. While I would argue against a mass rewilding of high-grade farmland, because of the threat this could present to global food supplies, we lose little by allowing nature to persist in small fallow corners and unexploited pockets of even the most fertile places.

The drive towards monoculture causes a dewilding, of both places and people. It strips the Earth of the diversity of life and natural structure to which human beings are drawn. It creates a dull world, a flat world, a world lacking in colour and variety, which enhances ecological boredom, narrows the scope of our lives, limits the range of our engagement with nature, pushes us towards a monoculture of the spirit.

I doubt that anyone wants this to happen to the land that surrounds them, except those – a small number – who make their money this way. But these few have been empowered both by their ownership of the land and by a kind of cultural cringe, which prevents other people from challenging them. The Italian philosopher Antonio Gramsci used the term 'cultural hegemony' to describe the way in which ideas and concepts which benefit a dominant class are universalized. They become norms, adopted whole and unexamined, which shape our thinking. Perhaps we suffer from agricultural hegemony: what is deemed to be good for farmers or landowners is deemed, without question or challenge, to be good for everyone.

In some cases we pay to support this hegemony and the monocultures it creates. Scores of billions of pounds of public money are spent each year to sustain the degradation of the natural world. In the United States, farm subsidies encourage the unvaried planting, across vast acreages, of corn. In Canada, subsidies for pulp and paper mills help to replace ancient forests with uniform plantations. Worse, perhaps, from the point of view of rewilding, is public spending which sustains monocultures in places which would otherwise be reclaimed by nature. This is what happens in the nation I am using as a case study of the monomania which blights many parts of the world. Here another monoculture has developed: a luxuriance, an infestation, a plague . . . of sheep.

I have an unhealthy obsession with sheep. It occupies many of my waking hours and haunts my dreams. I hate them. Perhaps I should clarify that statement. I hate not the animals themselves, which cannot be blamed for what they do, but their impact on both our ecology and our social history. Sheep are the primary reason – closely followed by grouse shooting and deer stalking – for the sad state of the British uplands. Partly as a result of their assaults, Wales now possesses less than one-third of the average forest cover of Europe.[1] Their husbandry is the greatest obstacle to the rewilding I would like to see.

To identify the sheep as an agent of destruction is little short of blasphemy. In England and Wales the animal appears to possess full diplomatic immunity. Its role in the dispossession of many of the people who once worked on the land, as the commons were enclosed by landlords hoping to profit from the wool trade, is largely forgotten. This is what Thomas More wrote in *Utopia*, published in 1516:

> Your sheep, that were wont to be so meek and tame and so small eaters, now, as I hear say, be become so great devourers, and so wild, that they eat up and swallow down the very men themselves. They consume, destroy, and devour whole fields, houses, and cities. For look in what parts of the realm doth grow the finest and therefore dearest wool, there noblemen and gentlemen, yea and certain abbots, holy men no doubt . . . leave no ground for tillage, they inclose all into pastures; they throw down houses; they pluck down towns, and leave nothing standing, but only the church to be made a sheep-house . . . the husbandmen be thrust out of their own, or else either by cunning and fraud, or by violent oppression they be put besides it, or by wrongs and injuries they be so wearied, that they be compelled to sell all: by one means therefore or by other, either by hook or crook they must needs depart away.[2]

In Scotland, where the Clearances were more sudden and even more brutal than the enclosures in England and Wales, some people remain aware of the dispossession and impoverishment caused by sheep farming. But in Wales, though sheep have replaced people since the Cistercians established the Strata Florida abbey in the twelfth century, and though these enclosures were bravely resisted by riots and revolts such as *Rhyfel y Sais Bach* (the War of the Little Englishmen) in what is now Ceredigion in 1820,[3] the white plague has become a symbol of nationhood, an emblem almost as sacred as Agnus Dei, the Lamb of God, 'which taketh away the sin of the world'. I have come across a similar fetishization in Australia and New Zealand, North America, Norway, the Alps and the Carpathians.

There is a reason for this sanctification, but it is rapidly becoming outdated. While sheep were used in Wales as an instrument of enclosure in the eighteenth and nineteenth centuries, during the twentieth there was a partial but widespread process of land reform in the uplands. In the aftermath of David Lloyd George's People's Budget of 1909, which increased income tax and inheritance tax for the very rich, the big landowners in Wales, many of whom were English, began to sell off some of their property.[4] They appear to have been less attached to their Welsh estates than to their English properties or their sporting land in Scotland, so these were shed first. Much of the land was bought by their tenants. Partly as a result, a smaller proportion of Wales than of England or Scotland remains in large estates. As the farmer Dafydd Morris-Jones, with whom I have discussed these issues at length, pointed out to me: 'there is a great sense of national pride in the fact that the local population, after centuries of subservience, were able to reclaim "their" lands, and were no longer beholden to the lord of the manor'.

After the Second World War, through the 1947 Agriculture Act and the 1948 Agricultural Holdings Act, the tenant farmers who continued to rent their land gained security for life. For eighty or ninety years, until quite recently, much of the land in Wales was controlled by small farmers, most of whom raised sheep and cattle. (The cattle gradually disappeared, partly, it seems, as a result of the loss of the suckler cow premium – a European subsidy – in 2003.) During a period in which it faced mortal threats, they sustained the Welsh language and important elements of the national culture. Now the family farms are consolidating rapidly, into new agricultural estates. Despite the £3.6 billion a year British people spend ostensibly to sustain a viable farm economy, the National Farmers' Union reports that '21% of upland farms are not expected to continue beyond the next 5 years.'[5] The brief flowering of small-scale farming appears to be coming to an end.

Until the enclosures, Welsh farmers kept large numbers of cattle and goats in the uplands, and grew cereals, root crops and hay, even, in some places, on the tops of the hills. By the end of the nineteenth century, and the coming of the railways, much of this mixed farming had been replaced by sheep and cattle. The enclosures consolidated a grazing culture which still resonates through the place names, ballads and oral traditions of Wales. Farmers moved their flocks between *hendre* – literally 'old town' (the winter grazings surrounding the farmstead) – and *hafod*, rough huts in the summer pastures on the hills, some of which eventually became solid stone houses. (I have seen a similar system in Transylvania, where, in the late 1990s, shepherds who rode fine black horses still slept in summer houses, or *stînas*, of sticks and shakes in the mountains, milked their sheep and cows in the pastures, made a white cheese which they hung in bags from the rafters, drank plum brandy and sang around the fire at night.) Drovers walked the sheep along ancient tracks into England, driving the flocks from the Welsh uplands to markets as distant as Kent. Shepherds bred dogs and trained them to perform astonishing feats. Most of this has now gone, or persists – in the form of sheepdog trials – as little more than a ghost of the economy it once served.

Subsidies after the Second World War encouraged the farmers to increase the size of their flocks. Between 1950 and 1999, the number of sheep in Wales rose from 3.8 to 11.6 million. After headage payments – grants for every animal a farmer kept – were stopped in 2003, the population fell back again, to 8.2 million by 2010,[6] which is still almost three sheep for every human being in Wales.

Since the Second World War, sheep have reduced what remained of the upland flora to stubble. In 6,000 years, domestic animals (alongside burning and clearing for crops and the cutting of trees for wood, bark and timber) transformed almost all the upland ecosystems of Britain from closed canopy forest to open forest, from open forest to scrub and from scrub to heath and long sward. In just sixty years, the greatly increased flocks in most of the upland areas of Britain completed the transformation: turning heath and prairie into something resembling a bowling green with contours.

Though sheep numbers have begun to decline, the impacts have not. More powerful machinery allows farmers to erase patches of scrub growing on land that was previously too steep to clear. This allows them to expand the area that qualifies for subsidies. In mid-Wales some farmers appear to retain a powerful compulsion, as they sometimes put it, to 'tidy up' the land. Ancient hawthorns and crab apples close to my home, often the last remnants of the last hedges on hills that are otherwise devoid of trees, are still being ripped up and burnt, for no agricultural reason that I can discern, except a desire for neatness and completion. From my kayak in Cardigan Bay I see a sight that Neolithic fishermen would have witnessed: towers of smoke rising from the hills as the farmers burn tracts of gorse and trees.

The UK's National Ecosystem Assessment shows that the catastrophic decline in farmland birds in Wales has accelerated, despite the reduction in the number of sheep: in the six years after 2003 their abundance fell by 15 per cent.[7] Curlews declined by 81 per cent in just thirteen years (from 1993) and lapwings by 77 per cent in only eleven years (from 1987). Golden plover, which have been the focus of intense conservation efforts, are now almost extinct: reduced to just thirty-six breeding pairs.[8] Even in the most strictly protected places, only 7 per cent of the animal and plant species living in rivers are thriving.[9]

Overwhelmingly the reason is farming: grazing which prevents woods from regenerating and destroys the places where animals and plants might live, the grubbing up of trees, cutting and burning, pesticides and fertilizers which kill wildlife and pollute the watercourses. Almost all the rivers in Wales are in poor ecological condition, which is unsurprising when you discover that the nitrates and phosphates entering the water have risen sharply.[10] Sheep dip residues have been found in almost 90 per cent of the places scientists have surveyed.[11] Sheep dip is especially damaging, as it contains a powerful pesticide – cypermethrin – which can kill much of the invertebrate life in a river. Farming is cited as a reason for the decline of wildlife in Wales in 92 per cent of cases.[12]

A similar story can be told in almost all the uplands of Britain: Dartmoor, Exmoor, the Black Mountains, the Brecon Beacons, Snowdonia, the Shropshire Hills, the Peak District, the Pennines, the Forest of Bowland, the Dales, the North York Moors, the Lake District, the Cheviots, the Southern Uplands. In fact the only wide tracts of upland Britain not grazed to the roots by sheep are those grazed to the roots by overstocked deer, in the Highlands and Islands of Scotland. Sheep farming in this country is a slow-burning ecological disaster, which has done more damage to the living systems of this country than either climate change or industrial pollution. Yet scarcely anyone seems to have noticed.

It grieves me to discover this. Hill farmers are trying only to survive, and theirs is a tough, thankless and precarious occupation. But when hills are heavily grazed – wherever in the world this takes place, – the other people of the nation pay a remarkably high ecological cost for this industry.

Those who defend heavy grazing – whether in Wales or Wyoming – sometimes argue that if sheep or other animals were removed from the hills, the ecological quality of the land would decline as trees and scrub replaced the grass. The National Farmers' Union of Scotland warns that 'fewer sheep ... means undergrazing of traditional pastures, a loss of biodiversity, a return to bracken and brash and the potential for irreparable damage to Scotland's beautiful landscape'.[13] The president of the Farmers' Union of Wales claims that reducing the number of sheep 'has a severe detrimental impact on upland biodiversity'.[14] This is incorrect. As I will show later, they appear to have confused a functioning ecosystem with a tidy one.

A more powerful argument is that upland grazing is essential for food production. This sounds likely, but is it really true? If Wales is a useful case study, perhaps not. Just over three-quarters of the area of Wales is devoted to livestock farming,* largely to produce meat.† But, by value, Wales imports seven times as much meat as it exports.[18] This remarkable fact suggests an astonishing failure of productivity.

That is not quite the end of the issue. Deep vegetation on the hills absorbs rain when it falls, and releases it gradually, delivering a steady supply of water to the lowlands. When trees and shrubs are removed, the rain flashes off the hills, causing floods downstream. Sheep also compact the topsoil, reducing its permeability, which ensures that still less water is absorbed. Drainage systems dug in the pastures accelerate these effects. When the floods abate, water levels fall rapidly. Upland grazing contributes to a cycle of flood and drought.

The results can be seen in the record of floods in the River Wye across the seventy years beginning in 1936.[19] The Wye rises on Pumlumon in the Cambrian Mountains. In this period the number of floods each year has

---

* The National Ecosystem Assessment states that 'agricultural land occupied some 1.64 million ha or 79% of Wales in 2008' and that 'crops now account for only 3% of the agricultural land area'.[15]

† Most of the animals farmed are sheep, whose major product is meat. There are also over 1 million cattle.[16] These are split almost evenly between dairy and beef,[17] but the male calves from both industries are reared for beef.

roughly tripled. Yet there has been no commensurate rise in rainfall.[20] Two things have changed. The first is that, as I have mentioned before, until the late 1990s the authorities dragged woody debris out of the upper reaches of this river, hastening the flow of water to the flooding zones. The second is that, as sheep numbers have risen, grazing in the watershed has intensified. Environmentalists have tended to blame all increased flooding on climate change. It is rapidly becoming a major factor, but until recently that was not the case. The land's reduced ability to absorb the water that falls on it appears to have been more important.

The rivers which drain the Welsh uplands, the Severn and Wye in particular, flow, when they reach the lowlands, through some of the most productive parts of Britain, where the soil is fertile enough to grow fruit and vegetables as well as cereal crops. Many of the farms here depend on irrigation. Many lose crops and opportunities when the land floods. It is not easy to estimate how much potential food production might be lost in such places as a result of the increased volatility of the rivers that pass through them, and I can find no research which attempts to do so. But, given the remarkably low output in the upland areas of Britain, it is within the range of possibility that hill farming creates a net loss of food. There must be few industries in which such extensive environmental damage supports such small gains and so few people.

Grazing is one of the least productive uses to which the hills could be put. Despite the vast area it occupies and the subsidies it receives, farming in Wales contributes just over £400 million to the economy.[21] Walking, with much lower environmental impacts, produces over £500 million, and 'wildlife-based activity' generates £1,900 million.*[22] The National Ecosystem Assessment shows that, across most of the uplands of Wales, switching from farming to multi-purpose woodland would produce an economic gain.[23] In other words, the current model of farming, far from being essential to the rural economy, appears to drag it down. The barren British uplands are a waste in two senses of the word.

All this would be less of our business if we were not paying for it. Hill farming is entirely dependent on subsidies provided by taxpayers. In Wales, the average subsidy for sheep farms on the hills is £53,000. Aver-

---

* This covers conservation work, wildlife tourism, other jobs which would not exist were it not for wildlife, and academic and commercial research and consultancy.

age net farm income is £33,000.[24] The contribution the farmer makes to his income by raising sheep and cattle, in other words, is minus £20,000.

Farm subsidies cost the United Kingdom £3.6 billion a year. They consume 43 per cent of the European budget: €55 billion, or £47 billion.[25] The British government estimates that the Common Agricultural Policy stings every household in the UK for £245 a year.[26] That is equivalent to five weeks of food for the average household,[27] or slightly less than it lays down in the form of savings and investments every year (£296).[28] Using our money to subsidize private business is a questionable policy at any time. When important public services are being cut for want of cash, it is even harder to justify.

What do we receive in return for this generosity? The Common Agricultural Policy raises the price of feed, chemicals and machinery, helping to drive the smaller farmers out of business. It raises the price of land, which excludes young people who want to become farmers, and contributes to the rising price of food. This vast expenditure of public funds supports remarkably few people: in the whole of Wales there are just 16,000 full-time and 28,000 part-time farmers.[29] But above all it pays for ecological destruction.

This is not an accident of policy. The rules are quite specific. They are laid down in a European code with the Orwellian title of 'Good Agricultural and Environmental Condition'. Among the compulsory standards it sets is 'avoiding the encroachment of unwanted vegetation on agricultural land'.[30] What this means is that if farmers want their money they must stop wild plants from returning.* They do not have to produce anything, to keep animals or to grow crops there; they merely have to prevent more than a handful of trees or shrubs from surviving there, which they can do by towing cutting gear over the land.

The infamous 'fifty trees' guideline ensures that pastures containing more than fifty trees per hectare are not eligible for funding. A survey by the Grasslands Trust found that this rule excludes farm habitats of great value to wildlife, such as the wooded meadows of Sweden, the limestone pavements of Estonia and the browsed scrubland of Corsica.[31] In Germany, pastures are disqualified from subsidies by the

---

* These conditions apply to Pillar 1 subsidies, which account for the majority of farm payments.

presence of small areas of reeds. In Bulgaria, the existence of a single stem of dog rose has rendered land ineligible. In Scotland farmers have been told that yellow flag irises, which for centuries have gilded the fields of the west coast, could be classed as 'encroaching vegetation', invalidating their subsidy claims. The government of Northern Ireland has been fined £64 million for (among other such offences) giving subsidy money to farms whose traditional hedgerows are too wide.[32] The effect of these rules has been to promote the frenzied clearance of habitats. The system could scarcely have been better designed to ensure that farmers seek out the remaining corners of land where wildlife still resides, and destroy them.

A farmer can graze his land to the roots, run his sheep in the woods, grub up the last lone trees, poison the rivers and still get his money. Some of the farms close to where I live do all of those things and never have their grants stopped. But one thing he is not allowed to do is what these rules call 'land abandonment', and what I call rewilding. The European Commission, without producing any evidence, insists that 'land abandonment in less advantageous areas would have negative environmental consequences'.[33]

To abandon is to forsake or desert. Abandonment is one of those terms – such as improvement, stewardship, neglect and undergrazing – which create the impression that the ecosystem cannot survive without us. But we do not improve the ecosystem by managing it; we merely change it. Across Europe, these rules have turned complex, diverse and fecund ecosystems into simple and largely empty ones. They have helped precipitate an ecological catastrophe.

There is a second tranche of subsidies that pays farmers to undo some of the damage inflicted by this system. It is a crazy use of public funds. First farmers are forced to destroy almost everything; then they can apply for a smaller amount of money to put some of it back.

But only a little. The 'green' subsidies (known as Pillar 2 payments) reward farmers for making marginal changes, and only in certain places. National governments disburse this money, using the European rules as their guidelines. The Welsh government assures farmers that these payments 'will require at most minor modifications to farming systems'.[34] In fact it expressly forbids them to restore more than a few tiny corners of their land. For example, the payment for allowing land 'to revert to rough grassland or scrub' applies only to areas of one-third of a hectare or less.[35] While the scheme provides subsidies for everything from the removal of coarse fish to the erection of kissing-gates,[36] there are no payments for planting native trees in most of the upland areas of Wales: tree-planting grants, on the whole, can be issued only for the lowlands and valleys, where the farmland is most productive and farmers are least inclined to use them.*

Farmers are supposed to prove that they have taken the measures for which they are receiving these 'green' payments. But enforcement falls somewhere on the spectrum between weak and hopeless. A friend whose job involved checking that farmers who are being paid to keep their sheep out of the woods are doing so tells me that 'the vast majority of farming schemes I checked failed, and represented what were basic-ally fraudulent claims'. He routinely found woods from which sheep were supposed to have been excluded full of the white plague, but when he recommended that the grant be stopped, the senior official at the time told him he must be mistaken, and that if there were a problem he should try merely to persuade the farmer to meet the conditions.

It seems puzzling, when subsidies have been removed from almost every other industry, that farming continues, despite the financial cri-sis, to receive so much support from taxpayers. I struggle to understand why there is not more public protest around this issue. Perhaps these payments – and the rules which govern them – reflect a deep-rooted fear of losing control over nature. We have not wholly shed our sense of a sacred duty to proclaim 'dominion over the fish of the sea, and over the fowl of the air, and over the cattle, and over all the earth, and over every creeping thing that creepeth upon the earth'.[38] But that may not be the only explanation.

'Charlemagne', writing in the *Economist*, has coined what he or she calls the 'Richard Scarry rule': 'Politicians will rarely challenge interests that feature in children's books.'[39] It is an appealing idea, though it does not seem to apply to other sectors: they willingly do battle with train drivers, for example. But perhaps it is relevant to farming. A large

---

* The Forestry Commission publishes maps which show where tree planting is and is not eligible for grants. It is beginning to ease the rules a little following widespread complaints about its discouragement of upland planting.

proportion of the books produced for very young children concern this industry. They tell a story of quaint and charming farmyards in which one cow and her calf, one sheep and her lamb, one hen and her chicks, one pony, one pig, one dog, one duck and one cat range freely. The farmers have broad smiles and rosy cheeks and live in arcadian peace with the animals they keep. Understandably, the issues of slaughter, butchery, consumption, castration, tusking, separation, battery production, farrowing crates, pesticides, waste disposal and other such industrial realities never feature. Unintentionally these books might implant, at the very onset of consciousness, a deep, unquestioned faith in the virtue and beauty of the farm economy and the importance of sustaining it, regardless of demand.

I spent several months pursuing an explanation for the subsidy rules, and the way they are interpreted by national governments, during which I was passed from one agency to another. After a long and exasperating correspondence with her civil servants, I secured an audience with the Welsh minister then in charge of rural affairs, Elin Jones. I began to understand the nature of the problem when she put down her file of notes on the table, and placed beside it a National Farmers' Union pen.

I was keen to discover why the Forestry Commission in Wales, a branch of the Welsh government, had issued a blanket ban on tree planting grants across almost all the uplands.* The explanation she gave astonished me: she claimed that allowing trees to return to the uplands would exacerbate global warming, as carbon dioxide would be released from the soil. When I asked her officials how this statement could be justified, they sent me two long scientific reports. I read them and discovered that they said the opposite of what the minister and her department had claimed. One of them revealed that it is not tree planting but overgrazing by sheep which has reduced the amount of carbon in the soil in the Welsh uplands.† Even plantation forestry, which cre-

---

* In 2011 the Forestry Commission published a map showing where grants will be issued for planting woodland. Almost all the upland areas of Wales, including most of the Cambrian Mountains, were marked red, meaning that no planting would be sanctioned there.[40]
† 'This has had a detrimental effect on the ranker and peaty podzol soils, with degraded areas containing significantly less carbon and nitrogen, means of 5% C and 0.4% N in comparison with 24–27% C and 1.1–1.4% N in intact heathland ecosystems at the same site.'[41]

ates much greater disturbance of the soil than allowing native trees to spread, causes no demonstrable carbon loss.[42] The other told me that in all the situations it modelled, planting trees on grasslands increased the amount of carbon in the soil.[43]

Yet Elin's argument is used across the European Union to prevent reforestation of the uplands. The European Commission claims that less farming would cause the 'loss of possibilities to contribute to the mitigation of climate change'.[44] It provides no evidence to support this statement. It would be highly surprising to discover that forest and scrub have a worse impact on the atmosphere than sheep or cattle farming.*

Subsidies are not the only means by which we pay for grazing in the hills. In England and Wales, floods cause around £1.25 billion of damage a year.† Protecting land and homes from possible impacts costs a further £570 million a year. The immediate reason for the summer floods that struck the region in which I live in 2012, flushing through houses, forcing the evacuation of the village of Pennal and the rescue by helicopter and lifeboat of campers and caravanners on the coast, drowning roads, railways and the electricity substation, was an Atlantic gale that dumped a very heavy load of rain on the hills.[47] But the floods must have been exacerbated – and might have been caused – by the reduced capacity of the hills to absorb this rain. Instead of percolating away slowly, it now sluices almost immediately into the valleys.

I am told by a senior civil servant that an insurance company recently investigated the possibility of buying and reforesting Pumlumon – the largest mountain in the Cambrians, on whose slopes both the Severn and the Wye arise. It had worked out that this would be cheaper than paying out for carpets in Gloucester. It abandoned the plan because of the likely political difficulties.

---

* Not only does the soil beneath woodland lock up more carbon than the soil beneath grass, but the trees also store more carbon above the surface: broadly speaking, trees are pillars of wet carbon. Sheep and cattle produce large quantities of methane, which is a powerful greenhouse gas. The tractors and quad bikes farmers use consume fossil fuels.
† 'The average annual cost of damage from flooding in England is estimated at more than £1 billion.'[45] The figure for Wales is, or was, £262 million. This is likely to have risen as a result of the floods in 2012.[46]

Strong as the case for change may be, agricultural hegemony is so potent that to challenge farmers and landowners is almost taboo. In Wales, farmers (both full- and part-time) account for 1.5 per cent of the total population and 5 per cent of the population of the countryside: 44,000 out of 960,000 *rural* people.[48] Yet the countryside is governed and managed almost exclusively for their benefit. Many of the ideas and perspectives which dominate rural policy arise with farmers' unions, which are often governed by the biggest and richest landowners. The views of the majority of rural people who are not farmers – 95 per cent in Wales – are marginalized. Elin Jones was minister for rural affairs, not minister for farming, but the pen she brought to our discussion was a cipher for her department's policies. Rural politics throughout Europe and in much of North America suffer from the same blight: their primary purpose appears to be to keep the farmers (or foresters or fisherfolk) happy, though everywhere they are a small minority.

I am convinced that this can change, that if people were more aware of how their money is being used, the needless destruction, the monomania, driven by farm subsidies – across Europe and in several other parts of the world – would come to an end. This, more than any other measure, would permit the trees to grow, bring the songbirds back, prompt the gradual recolonization of nature, release the ecological processes that have been suppressed for so long. In other words, it would allow a partial rewilding of the land.

10

The Hushings

. . . the smashed faces
Of the farms with the stone trickle
Of their tears down the hills' side.
R. S. Thomas
Reservoirs

Of all the world's creatures, perhaps those in the greatest need of rewilding are our children. The collapse of children's engagement with nature has been even faster than the collapse of the natural world. In the turning of one generation, the outdoor life in which many of us were immersed has gone. Since the 1970s the area in which children may roam without supervision in the UK has decreased by almost 90 per cent, while the proportion of children regularly playing in wild places has fallen from over half to fewer than one in ten.[1]

Parents are wrongly terrified of strangers and rightly terrified of traffic. The ecosystem of the indoor world has become ever richer and more engaging. In some countries, children are now demonized and harried when they gather in public places; their games forbidden, their very presence perceived as a threat.[2] But as Jay Griffiths records in her remarkable book *Kith*, they have also been excluded from the fortifying commons by the enclosure and destruction of the natural world.

The commons was home for boy or bird but the Enclosures* stole the nests of both, reaved children of the site of their childhood, robbed

---

* Enclosure, the worldwide process of privatizing or in some cases nationalizing common land, excluding the people and the uses to which it had formerly been put, was

them of animal-tutors and river-mentors and stole their deep dream-shelters. The great outdoors was fenced off and marked 'Trespassers Will be Prosecuted.' Over the generations, as the outdoors shrank, the indoor world enlarged in importance.[3]

As Griffiths shows, enclosure, accompanied by a rapid replacement of the commoners' polyculture with a landlord's monoculture, destroyed much of what made the land delightful to children – the ancient trees and unploughed dells, the ponds and rushy meadows, the woods, heath and scrub – and banned them from what it failed to destroy. Destruction and exclusion have continued long beyond the nineteenth century. So many fences are raised to shut us out that eventually they shut us in.

Enclosure, Griffiths notes, also terminated the long cycle of festivals and carnivals through which people celebrated their marriage to the land, when authority was subverted and mischief made. The places where the festivals had been held were closed, fenced and policed.

In the early 1990s, I saw this excision performed with shocking speed in Maasailand. I watched the warriors of the community with which I worked perform their people's last ceremonies – last rites – as the commons in which these had been held were privatized and wired up.[4] This process of enclosure and closure shut the people out of their land almost overnight, shattered their communities, dispersed their peculiar culture and drove the young people, many of whom were now destitute, into the cities, where their contact with the natural world was permanently severed. I watched, in other words, the recapitulation of the story of my own land, and witnessed the bewilderment, dewilderment and grief it caused.

The commons belonged, inasmuch as they belonged to anyone, to children. Their trees and topography provided, uncommissioned and unbuilt, the slides and climbing frames, sandpits and ramps, seesaws and swings, Wendy houses and hiding places which must now be constructed and tested and assessed and inspected, at great expense and (being planned and tidy, fenced and supervised) one-tenth of the fun. Their sticks and flowers and insects and frogs were all the toys that

consolidated and accelerated in England in the eighteenth and nineteenth centuries by parliamentary Acts of Enclosure.

children needed to fill their world with stories. 'Childhood,' Griffiths tells us, 'was to be enclosed as surely as the land.'

The impacts have been pernicious, but they are so familiar that we scarcely see them any more. The indoor world is far more dangerous than the outdoor world of which parents are so frightened, the almost non-existent stranger danger replaced by a real and insidious estrangement danger. Children, confined to their homes, become estranged from each other and from nature. Obesity, rickets, asthma, myopia, the decline in heart and lung function all appear to be associated with the sedentary indoor life.

Some studies, summarized in Richard Louv's book *Last Child in the Woods*, appear to link a lack of contact with the natural world to an increase in attention deficit hyperactivity disorder.[5] Research conducted at the University of Illinois suggests that playing among trees and grass is associated with a reduction in indications of ADHD, while playing indoors or on tarmac appears to increase them.[6] One paper suggests that playing out of doors improves children's reasoning and observation,[7] another that outdoor education enhances their reading, writing, science and maths.[8] Perhaps children would do better at school if they spent less time in the classroom.

Missing from children's lives more than almost anything else is time in the woods. Watching my child and others, it seems to me that deep cover encourages deep play, that big trees, an understorey mazed by fallen trunks and shrubs which conceal dells and banks and holes and overhangs, draw children out of the known world and into others. Almost immediately the woods become peopled with other beings, become the setting for rhapsodic myth and saga, translate the children into characters in an ageless epic, always new, always the same. Here, genetic memories reawaken, ancient impulses are unearthed, age-old patterns of play and discovery recited.

One difference between indoor entertainment and outdoor play is that the outdoors has an endless capacity to surprise. Its joys are unscripted, its discoveries your own. The thought that most of our children will never be startled by a dolphin breaching, a nightingale simging, the explosive flight of a woodcock, the rustle of an adder is almost as sad as the disappearance of such species from many of the places in which we once played.

I would like to see every school take its pupils, for one afternoon a week, to run wild in the woods. But there is a major hindrance: not enough woods. Many urban children live so far from the nearest woodland that this simple venture would entail a major expedition. Could every new housing development include some self-willed land in which children can freely play?

Even beyond the cities, in many parts of the world the woods have been erased. But now that farming, in the absence of subsidies, has become unviable in certain places, we could be about to witness the reversal of some of the enclosures which have excluded children and adults, and the wildlife in which we once exulted.

I recognize that there are conflicts here, that the vision I have begun to adumbrate in this book collides with other people's visions. The details differ in every nation, but the story is more or less the same: forms of farming or fishing or forestry which suppress the natural world are seen by those who pursue them as essential to maintaining the economy, culture and traditions of their communities. I have seen such struggles ignite loggers and fisherfolk in Canada, farmers in Norway, whalers in Japan. The conflicts are real and cannot be lightly dismissed. What I am about to describe is particular to Wales, but in essence almost universal. It is a clash between the valid concerns of those who now own or use the land and the valid concerns of those who would like to re-engage with it, but currently find no purchase there.

St David's Day. *Dydd Gŵyl Dewi*. The buds of the sallows were about to break. The silk straining at the bracts was stretched so fine that they gleamed like beads of mercury. The twigs of the birches had turned mauve as the sap rose into them. Daffodils had risen from the ground on the verges, and now their pregnant buds swayed on stiff stems as the lorries swept past. Otherwise, from the road, there was no sign that spring was soon to break out of winter's prison. The pastures still slumbered in their hibernal colours, yellow and tan. Last year's bracken, now a deep, snow-trampled russet, clung to the mountains. The higher peaks – Cadair Idris, Aran Fawddwy, Tarren Hendre – were still dressed in skewbald motley: the dead grass appeared browner and darker beside the patches of glaring white.

The low sun was so bright and the shadows so crisp that the land looked as if it had been lit for a film. This would be the fourth consecutive year in which the customary British weather had been reversed: easterly winds, warm days and crisp nights in the spring, smeary, rain-lashed summers, still, warm autumns.

In the heart of the Cambrian Mountains, I drove up a bumpy track to a small stone farmhouse. In the green fields around it grazed Welsh speckle-faced sheep, with panda bear eyes and comical black noses. Clear water poured over a sill into a raised pool beside the tidy farmyard. A white and caramel sheepdog lunged and barked on the end of its chain.

Dafydd Morris-Jones and his mother, Delyth, came out to greet me. I had expected a much older man: he was still in his twenties. He had blue eyes, a handsome, open face, two earrings in the top of one ear and – appropriately for a sheep farmer – mutton-chop sideburns. Delyth had the same bright eyes. Her white hair came down to her shoulders. She looked fit and strong.

I had found Dafydd after writing to the Cambrian Mountains Society, to express my concern about its portrayal of the ecology and landscape of the plateau. It had passed my letter to him. Though I disagreed with some of what he wrote, I had been impressed by his clear reasoning and the breadth of his knowledge, so I had asked to meet him.

Delyth herded me into the house and sat me down in her little parlour. A Welsh dresser displaying her best crockery filled one wall. It had been nailed up by Dafydd's great-grandfather, she told me, after his son – Dafydd's grandfather – had, as a small boy, tried to climb it and had brought it down, smashing all the plates.

Their family had taken the tenancy of this farm in 1885, and had bought the land in 1942. Dafydd had just replaced the roof of one of his barns – that had held since the beginning of his great-grandfather's tenancy – using the original slate. 'It should do for the next 150 years,' he told me.

After tea and scones, he took me out onto his land. His sheep, which were beginning to swell, were still in the low pastures surrounding the house. Dafydd explained that he puts the ewes with the rams later than most farms do, so that they could lamb in the fields, rather than indoors. 'In the fields, the sheep don't give birth after dusk. If you do it in the shed, it's round the clock. But it's essential to get up early, as they start lambing at dawn. The crows line up on the fence, waiting for their moment. They'll pluck the eyes out of the lambs even before they're fully born. You have to be there to keep them off.'

As we walked up the track which cut across his land, I started to become aware that I was in the presence of an excellent mind. Over the next few hours, he would speak about the best way to rebuild a cheap hydroelectric turbine, the long-distance signalling system used by the Romans, the problems associated with acid waste lagoons in China, new caving routes through the disused slate mines, the difference between a clacker wheel and an overshoot wheel and a dozen other subjects, in every case with an unusual combination of lightness and authority. He had also prepared himself well for my visit: he had read and considered the key texts on the subject I had come to discuss. He was – and this is a word I seldom use – a brilliant young man. He could have done anything. But he had chosen the sparsest and hardest of livings. It also became clear to me that he had something else few people possessed: he knew who he was. I envied him that.

Dafydd had a degree in Welsh, from Cardiff University. He spent half his time farming and divided the rest between translation work (mostly in the winter) and outdoor education (mostly in the summer). He was deeply embedded in the life of his valley, helping to run, for example, the community woodland that had replaced a local conifer plantation. 'Here,' he told me, 'you've got the history of the nation written out in the landscape.'

The low sunshine exposed every scratch and tump of the sheep-shaved ground. Half-buried in turf were the remains of a drystone wall – first built, Dafydd said, in 1680 – that once separated the two great estates whose boundary his farm had straddled. It ran across the many miles of moor and mountain from Pumlumon to Cwmystwyth. Half of one of the estates had been lost – as tradition demanded – in a card game, which was why the farm whose tenancy his great-grandfather had later taken had been split between two owners. Among the knolls and tummocks he pointed to were Bronze Age burial mounds, medieval longhouse platforms and mystery enclosures, which might have been fishponds, but appeared to be in the wrong place. The low nobbly hill facing us, he told me, belonged to a farm which was mentioned in the *Mabinogion*, Welsh legends some 1,500 years old.

Beside the path was a pile of stones, sketching the barest outline of four walls, now sinking back into the close-cropped grass. 'That house was last inhabited in 1916, by the old cook who worked at my mum's school.'

I followed him up the side of the hill to a patch of brighter grass and soft-centred rush. This, he said, was the remains of an old hushings. It was either Roman or medieval: the archaeologists had not been able to decide. I confessed that I did not know what the word meant.

He explained that it was part of the valley's old lead-mining system. The miners built a dam above the deposits they wanted to expose, and channelled water through a leat into the pond it held. When the reservoir was full, they would breach the dam and the water would rush down the hillside, sweeping away the overburden. This was, in other words, the method I had seen deployed in the goldmines of Roraima, but without the use of diesel pumps.

Both the grass and the land it covered became rougher as we climbed. Dafydd explained that, to obtain green subsidies,* he had to keep his sheep off the mountain in the winter. He led me up into his summer grazings on Mynedd yr Ychen, Oxen Mountain. Short tufts of heather, still in the black mourning clothes of winter, survived amid the grass. Last year's dried flowers rattled on the stems. As we reached the crest of the hill, the great yellow plateau opened up. It rose towards that least distinct of mountains, Pumlumon Fawr, which, upwelling gently from the massif, always looks smaller than it is. Its grey and yellow flanks were patched with artless blocks of spruce. But for the wind, the land was silent. As usual in the Cambrians, no birds called and nothing rustled in the grass.

The heather on this pasture, Dafydd told me, might explain the name of the mountain, as cattle need a large amount of copper in their diet, and heather is a rich source. That it was called Oxen – not Cattle – Mountain suggested that the name pre-dated the era of horse

* Pillar 2 payments.

traction: oxen were used for heavy work from the Bronze Age until a few centuries ago. The boundary wall between the mountain and the winter pastures closer to the house, he later told me, had been built to allow the sheep to regulate their own grazing. It was banked up on the downhill side, to allow the sheep to move onto the mountain when their grazing in the lower fields declined, but not on the upper side, to ensure that they could not return until the farmer wished it.

Clinging to the hillside below us were the crumbled walls of a small stone building. 'That was the old goose house. Grandma used to walk up here every night to shut them in. The geese grazed on the grass and heather tips. The farming was more mixed in the past. Until 2000 we had a small herd of Herefords, which had come down from my great-grandfather's cows.'

Dafydd pointed out where the old farmsteads of his neighbours had stood, in some cases just three or four decades before. 'At night there were lights twinkling all along the valley. Now they've gone.'

He explained that this valley was once a busy thoroughfare. It was used by people walking to the church, to school and to the pub, which had now closed. It was used by the pilgrims who arrived at the docks in Aberystwyth (which were demolished long ago) to walk to the Cistercian abbey at Strata Florida. It was used by the drovers herding animals along the old trails to Rhayader and then to London.

'Our history is carried by word of mouth, but it's anchored to the land. The old boys used to play a game: one of them would leave his cap on a rock, somewhere in the mountains. Then he'd go into the pub and tell the name of the rock to a friend. That was all the information they needed. The friend had to run out and retrieve it. All the rocks had names. My uncle could remember all of them. They were never written down.'

Listening to him, I realized that both of us were harking back to something that is no longer here. His thoughts were filled by the days in which the hills bustled with human life. Mine were filled by the days in which they bustled with wildlife.

We came down the western side of the mountain, through low tussocks of gorse and heather, into the greener fields behind his house. As we approached the farmyard gate, we met Delyth, driving up the hill towards us on a quad bike with a trailer of hay, her white hair flying in the wind. She looked like Boudicca on her chariot. 'I hope you'll stop for lunch. It's ready now,' she said.

'I'm trying to limit the physical work she does,' Dafydd told me. 'But farming's in her blood and you can't stop her. She's only rolled the bike four times.'

When she had fed the sheep, Delyth ushered us into her parlour again and served us *cawl* made from one of her own small flock of turkeys, sweetened with swede and carrot, and brown bread, still warm from the oven. She and Dafydd began to tell me about the history of their farm and the community.

They explained that the estates started to form in the 1640s. The people here were not cleared, but had to pay to remain on the land they were already farming and had long seen as theirs. The first landlords were members of a Welsh aristocracy – Prices, Vaughans, Johneses – families which had supported Owain Glyndwr's uprising. They helped to keep the culture and language alive. At Hafod Uchtryd, the great estate which had owned the eastern half of the farm, the Johneses kept a Welsh printing press in the cellar.

In 1833, the Duke of Newcastle took over the estate. Delyth explained that attending the Anglican church rather than the Methodist chapel was a condition of tenancy: if you disobeyed the rule, you lost the farm. 'Dafydd's great-grandfather was worshipping and reciting in a language he didn't understand. But his great-grandmother insisted on going to chapel: she wouldn't speak to her Lord in English. It terrified her husband: we could have lost everything.'

'Our knowledge was not valued,' Delyth went on. 'The story was that people who stayed on the farms were the dimmest of all – so their knowledge must be dim as well. No one thought of writing it down. My father hardly wrote. He had to remember all the sheep figures, the prices and everything. There's not the same need to use our brains now.'

Dafydd was teaching himself the old Welsh counting system. Based on multiples of 10, 15 and 20, it was designed by shepherds for counting animals. 'You can juggle the numbers between the fingers of your two hands, totting the blocks on one, the individuals on the other. It allows you to count very quickly. In the new numbers, you can't count fast enough to match the speed that sheep run at. So you have to slow them down through the gate. From the 1970s onwards, Welsh learners were taught the decimal system. I can see the sense in that, but we've also lost something.'

He could, Delyth told me, judge the weight of a sheep to within a kilo; they had stopped using the scales as he always got it right, and it was quicker to weigh them by eye. She could do something he could not: she could spot their diseases at a distance, diagnosing them from the way the sheep stand or lie. She also knows just when they are going to lamb.

Dafydd gently moved the conversation onto the subject that divided us.

'My concern with rewilding is that it takes the people out. I see it as a post-Romantic ideology which imagines what the land would be like if only people weren't here. Look at what the Wildland Network* says at the bottom of its website: it wants the landscape "to be freed of human interference and managed with minimal intervention". That yells "cleansing" to me.'

There was, he explained, a deep local hostility to planting trees, as a result of the vandalism inflicted by the Forestry Commission during the middle decades of the twentieth century. As I had seen elsewhere in Wales, the commission launched a kind of Cultural Revolution in Wales, in which its green guards requisitioned ancient halls and farmsteads and dynamited them. In some cases they erased derelict villages,[9] and replaced them with party-approved plantations of identical Sitka spruce trees. It was a crime for which there has been little acknowledgement and no reckoning.

'The people of Myherin [the valley of the little stream to the east of his farm] were forced to leave: their land and homes were purchased under pressure. The commission planted 17,000 hectares of spruce where they had lived. Of the ten houses it bought, just three are still visible: two are in ruins, one is a bothy. The rest have just disappeared beneath the trees. The roots smashed up what remained. They destroyed all traces of the community.

'I'm not against something new, not by any means, but it should be a progression from what you've got, not wiping the slate clean. With

---

* This, though it shares the name of an active North American organization, was a British group, now either dormant or dead.

blanket rewilding you lose your unwritten history, your sense of self and your sense of place. It's like book-burning. Books aren't written about people like us. If you eradicate the evidence of our presence on the land, if you undermine the core economies that support the Welsh-speaking population in the language's heartland, you write us out of the story. We've got nothing else.

'Conservation should be about how we can live in nature. When it deviates from that, you forget that you're still looking at it from a human perspective. I think rewilding is an oxymoron. As William Cronon points out, if you argue for wilderness for its own sake, you're still imposing a human point of view.[10]

'People say they want to reintroduce predators. Why? The wolves don't miss being here. We'd be introducing them for the sake of alleviating human guilt about what we have done to the environment. Which is to meet a human need, not a wild need. It's all based on our own value judgements. I see rewilding as post-Romantic gardening. It's like those big rococo mansions with their toy milkmaid parlours.

'I'd much prefer to see trees here than wind turbines. But neither would keep the school open, support the local shop or reopen the pub. The average age of farmers in the UK is now sixty-two. It rises every year. The danger is that we have old people who speak the "old" language and a place barren of everyone else. That's a chilling thought.'

'It's also the visual impact,' his mother added. 'Without trees you can see all the lights from the other farmsteads across the valley. You don't feel so lonely. The forestry shuts us off from each other. It would bring despair with it if you're not careful.'

I found these arguments compelling, and I left the farm feeling troubled and confused. Two sets of values, both of which I held strongly, were fighting each other. I was painfully aware of the damage sheep have done to the upland ecology of Britain, and to the upland ecology of many other parts of the world. The bird surveys and other evidence suggest that the impacts are intensifying. The industry that causes this damage depends upon public subsidies, here and in many other countries. So we are paying both to sustain its assault on nature and to prevent the land and its ecosystems from recovering.

Yet the idea that Dafydd and Delyth and people like them should be pushed aside to make way for wildlife was also intolerable. I did not want to see their history erased or their culture blotted out, to witness a hushing: a sweeping away of the accumulated strata of their lives, a silencing of their voices.

I did have responses to some of the specific points Dafydd raised. The land and its economy have changed drastically over the past half-century. Much of the public money which would once have supported people like Dafydd and Delyth is now taken by ranchers, people who don't live on the land they farm and visit only when they have to. You can see abundant evidence of this long-distance farming on the roads of mid-Wales: Land Rovers driving this way and that, towing quad bikes in their trailers. The people who have bought this land are likely to have less interest in its history and culture. They are piggybacking on the moral capital of the Dafydds and Delyths, whose survival, for many taxpayers, is the only remaining justification for the extravagance of subsidies.

As absentee ranching spreads and mechanization advances, employment on the farms declines, as it is doing worldwide. Farming in Wales now produces less than a quarter of the income generated by wildlife, despite the fact that it occupies a much greater area than the land set aside for nature. I have yet to see any plan for hill farming which predicts that sheep raising will provide a growing or even stable share of national employment. The remaining farmers, like Dafydd, survive by making much of their income from activities other than farming. Rewilding, on the other hand, has great potential to attract walkers and nature-lovers. Though the Cambrian Mountains are close to the conurbations of the West Midlands, they are scarcely visited today.

In the early years, rewilding requires plenty of labour: planting trees, reintroducing lost plants and animals, removing fences and controlling exotic invasive species, such as rhododendron and Sitka spruce, and stray sheep. As the ecosystem recovered, the rewilding workforce would decline, but the potential for generating money from tourism would rise. Banishing the sheep and banishing the people are not the same thing. It is possible to envisage a thriving community of former farmers acting as wardens and guides, providing bed and breakfast, farm shops, clay-pigeon shooting, bicycle hire, horse riding, fishing lakes, falconry, archery and all the other services that now help rural communities to survive.

Researchers in North America have studied places in which extractive industries have given way to wildlife, with mixed conclusions. One paper, for example, states that 'employment and personal income levels in "wilderness" counties grew faster than in "resource-extraction" counties'.[11] Another maintains that in regions where timber cutting had stopped in order to protect the forests, economic wellbeing 'improved in some, deteriorated in some, and showed little change in other communities'.[12] The results are likely to be different in other nations, and the potential impacts, both positive and negative, should be carefully assessed. But it is possible that rewilding could do more than sheep farming to keep the school open, support the local shop or reopen the pub, which the current economy has manifestly failed to sustain.

As for book-burning, I see it whenever I walk in the hills close to where I live. I see oak woods, which in some cases had been preserved by farmers or mining communities for centuries, being destroyed by the sheep that now graze beneath the trees. I see hedgerows being grubbed up, drystone walls replaced with wire fences, ancient trees which once marked the boundaries between farms ripped out and burnt. Yes, rewilding could present a threat to the cultural history of the land. But I also see farmers from the communities which claim to treasure this history obliterating it, with scarcely a voice of protest raised against them.

If rewilding took place it would happen in order to meet human needs, not the needs of the ecosystem. That, for me, is the point of it. Wolves would be introduced not for the sake of wolves but for the sake of people. If rewilding happens it will be because we value a biologically rich environment more than we value an impoverished system which continues, with the help of public money, to support sheep.

After I showed Dafydd the first draft of this chapter, he responded to my suggestion that the people of this nation should decide whether wolves are introduced as follows:

Firstly, which people and which Nation? The loudest? The most well educated? The greatest percentage of the overall population? There is another value judgement here, do we value the enhancement and enrichment of outsiders' lives over the needs of the existing community, placing the recreational and emotional needs of for example West Midlanders over those of the local population? Isn't this the same argument that was used to further the cause of reservoir building (e.g. Liverpool's need for water in the case of Tryweryn) land clearance (our nation's need for defence training at Eppynt and Penyberth) and the Forestry Commission's afforestation (our growing nation's need for timber)?!

Surely, though, lamb is not produced to feed the farmers, but for sale to outsiders for the enhancement and enrichment of their lives. Changing the use of the land but not its ownership does not alter this relationship. But expropriation and dispossession of the kind deployed by the foresters, the reservoir builders and the army is a different matter. I would oppose any proposal to wrest land out of the hands of farmers for the purpose of rewilding. If rewilding is to happen, it must do so with the consent and involvement of those who currently work there.

But none of this is to dismiss the core argument which he and Delyth made so powerfully, and with which I find myself strongly in sympathy. They see rewilding as completing the long process of economic change and exclusion that has been erasing them and their culture from the land.

I found myself tumbling into cognitive dissonance, the uncomfortable state of mind that results from an inability to resolve conflicting ideas or values. I was unable to deny either position, yet each was exclusive of the other: I could not simultaneously support rewilding and the restoration of the ecosystem *and* support efforts to sustain the sheep farming that kept Dafydd, Delyth and their culture alive. I saw destruction and sadness in both directions. That is the sorry state in which I remained for several weeks.

Then, walking up the hill behind my house one morning, past a rare stand of birches that has recolonized a patch of rough grazing, the answer struck me. It was so simple, so obvious that I could not understand why I had failed to see it before.

As I mentioned earlier, sheep farmers in the Welsh hills receive an average of £53,000 a year in subsidies while their average net farm income is £33,000. Keeping livestock, in other words, costs them

£20,000 a year, though this gap may diminish if the price of lamb continues to rise. But, under the Common Agricultural Policy, if you want your subsidy payment, one of the few things you are forbidden to do is nothing. The Good Agricultural and Environmental Condition rules specify that if you do not keep the land clear, you forfeit everything. There is no requirement to produce anything; you must merely stop the land from reverting to nature, by either ploughing it, grazing it or simply cutting the resurgent vegetation. The purpose is to prevent the restoration of the ecosystem.

So here, perhaps, is the resolution of the conundrum that caused me such trouble: this rule should be dropped. Those farmers who are in it only for the money would quickly discover that they would earn more by lying on a beach than by chasing sheep over rain-sodden hills. Those who, like Dafydd and Delyth, believe in what they are doing, and have wider aims than just the maximization of profit, would keep farming. Where the life and community associated with raising sheep are highly valued, farming will continue. Where they are not, it will stop. Large areas of land would be rewilded, and the farmers who owned it could receive, as well as their main payments, genuinely green subsidies for the planting, reintroductions and other tasks required to permit a functioning ecosystem to recover. The alternative is the system we have at present: compulsory farming, enforced by the subsidy regime.

There is, I think, a necessary refinement of this simple idea. At present the subsidy system is deeply regressive. While it is funded by the taxes extracted from everyone, rich and poor, the money is disproportionately harvested by the biggest landowners. This, under the current system, is inevitable, as farmers are paid according to their acreage. According to Kevin Cahill, the author of *Who Owns Britain*, 69 per cent of the land here is owned by 0.6 per cent of the population.[13] It is profoundly wrong, I believe, that people struggling to support their families should be forced to extend alms to dukes, sheikhs and sharks: the absentee landlords, speculators and assorted millionaires who own much of the farmland of Britain and other parts of Europe.

To address this injustice, I would like to see the European Union introduce a maximum entitlement for the main subsidy payment.* I would suggest that no more than 100 hectares owned by a farmer, business or trust be eligible for this money. This would save a great deal of public funds while giving small farms (which are more labour-intensive) an advantage over the large ones. It could help to reverse the growing concentration of land ownership.

In renegotiating the Common Agricultural Policy, which governs the payment of subsidies, the Westminster government argued against any such cap, on the grounds that it would discourage 'consolidation' which, it says, enhances competitiveness.[14] In other words, the government wishes to see a greater concentration of ownership.

Dafydd points out that removing the obligation on landowners to farm their land could make ownership attractive to absentees, pushing up the price and squeezing farmers out of the market. This is a genuine danger, though it has very different consequences for the farmers who own their land (and could benefit from rising prices) and those who rent their land and might wish to exercise their right to buy it.

But the current subsidy system exerts the same effect: artificially inflating the price of land at the expense of tenants and new entrants (people who want to become farmers). It is hard to conceive of a subsidy system which would do otherwise: if there are to be farm payments of any description, they will cause the price of land to rise. The imposition of a cap would counteract this to some extent, making the land and the money to be harvested from it less attractive to the very rich.

These suggestions are ambitious. But something has to change. Economically, politically and ecologically, the current subsidy system is unsustainable. Eventually, all over Europe, it will break. We should prepare ourselves for this moment by developing a clear alternative. Far from being coercive, removing the abandonment rule would do nothing but enhance farmers' freedom of action, or freedom of inaction. Taxpayers would no longer find themselves obliged to fund just one vision of how the countryside develops. We would be paying for nature in some places, culture in others, and, except for sites of

---

* Namely, Pillar 1 subsidies or the single farm payment.

particular ecological importance, there is no need for anyone to specify where those places would be.

The farmers' freedom would create the space for other people's. Where they decide to stop cutting or grazing or burning their land, change could happen very quickly. Land which now supports the barest remnants of life, which is silent but for the wind and the sheep, would (with a little help at the beginning) soon become recolonized by trees and birds and insects, as Ritchie had discovered in one of the least auspicious corners of the Cambrian Desert. As the returning eco-system developed, some places would revert to deep forest, others, at first, to gorse and heath, others to carr: bog forest dominated by alders or willows or aspens. If we could then begin to reintroduce missing species – the large mammals absent for so long from these hills – places which nurture almost nothing but crows and tormentil could become as rich in life as some of the world's most famous national parks.

People as well as wildlife could regain a footing on the land. Tracts which have been reduced to a repellent bleakness, where there is no living structure, no natural shelter, could again exhilarate and entrance. Where there was little but brown grass before, where the exploration and discovery of nature end almost as soon as they begin, ecosystems could flourish which again beguile both children and adults, which offer endless adventures of revelation and surprise. I hope that at least some of these rewilded places will be big enough to prove uncrossable in one day's walk. A sense of boundlessness is something whose absence afflicts many rich nations. When, after half an hour walking across a wood, I reach the fence that separates it from the surrounding fields, I feel that something which was just beginning – a deep abstraction – is prematurely truncated. The dis-covery and wonder, the freedom from structured thought which had begun to open my mind come to an abrupt end.

In some parts of the world tumultuous nature is already returning to places from which it had been banished. One estimate suggests that two-thirds of those parts of the United States which were once for-ested, then cleared, have become forested again, as farming and logging have retreated, especially from the eastern half of the coun-try.[15] Another proposes that by 2030, even without any change in the subsidy regime, farmers on the European continent (though not in Britain, where no major shift is expected) will vacate around 30 million hectares of land, an area roughly the size of Poland.[16] This is not the result of any policy or plan; in fact, some European governments are trying very hard to stop it from happening and to keep farmers on the land. But as young people leave to find jobs and adventure elsewhere and no one is prepared to take their place, the decline of farming in many parts becomes inevitable.

There is a sadness here, which I felt while walking in the Ardèche in southern France, and finding, like Mayan ruins in the jungle, exquisitely built stone terraces, flagged paths, ancient bridges and stone stairways now overwhelmed by chestnut forests – growing sometimes from the very walls – through which sounders of boar marauded and pine martens leapt. My delight in the resurgent wildlife was tempered by the shock of seeing that work, laid down hand upon hand by untold generations, whose people – like Dafydd and his roof – had built a future for descendants they would never meet, gone all to waste. A civilization had been erased.

The process of retreat, with its mingled griefs and joys, appears in many places, particularly the uplands of Europe, to be inexorable. Unless farmers and their children are to be forced to remain on the land, there is no option but to acknowledge it and then to decide what happens next. The areas farmers will vacate might be large enough, if the people of this continent so choose, to permit the reintroduction not just of the wolves, bears, lynx and bison which are gradually regaining their footing on the land, but also of elephants, rhinos, hippopotamuses, lions and hyenas.

Does that sound ridiculous? I am sure it does. It is fair to say that the people of Europe are not yet ready for it. But if there is sufficient land, if that land is concentrated in large enough blocks and protected from further exploitation, there are likely to be few biological impediments. All these animals (or those of related species) ranged across Europe until recently, and our native fauna and flora have evolved to survive their attentions. The barriers, of course, would be political and cultural. But as the remarkable change in attitudes towards the wolf in many parts of Europe demonstrates, this might not always be so. Perhaps one day big cats will no longer need to be imagined.

As nature retreats from other parts of the world, Europe, the first continent to lose its megafauna and much of its mesofauna (the middle-sized animals), could, through rewilding, become one of the most biologically wealthy regions on earth. The story we have missed, while rightly lamenting the shocking collapse of biodiversity in so many countries, is that we could be about to witness a raucous European summer.

11

The Beast Within
(Or How Not to Rewild)

And I think in this empty world there was room for me and a
mountain lion.
And I think in the world beyond, how easily we might spare a
million or two humans
And never miss them.
Yet what a gap in the world, the missing white frost-face of
that slim yellow mountain lion!

D. H. Lawrence
Mountain Lion

Four Czech skinheads, dressed in black muscle shirts and combat
trousers, eyes glittering, jabbed their fingers at the weapons and talked
in low, intense tones. They strained with anger and excitement. For
them, it seemed, the war deemed to have finished almost a century
ago was not yet over. Here, 600,000 men had died in the First World
War, on a front now largely forgotten in northern Europe – the *Soška
fronta* – where soldiers of the Italian and Austro-Hungarian armies
faced each other in conditions as brutal and lethal as those on the
Somme, along the Soča valley and over the mountains, in some cases
across a few metres of bare peak, in trenches hacked into rock and ice.

Walking in the Julian Alps, we had followed the old supply lines,
seen concrete emplacements and stopped at the remains of cable stations which were used to haul equipment from one peak to another.
As we passed other hikers in bright colours, with friendly greetings in
a dozen languages, watched the ibex placidly chewing the cud in the
high mountain pastures and fed the choughs on scraps of cheese, the

horrors of that front were unimaginable. But here in the Kobarid Museum, the cases, the maps and panels began to make sense of what we had seen, and of the astonishing scale of the slaughter.

But as the skinheads hissed and whistled and clenched their teeth while they pored over the faded photographs, my partner pointed to something I had missed. As soon as she did so, I was riveted. Most of the panels showed the same thing, regardless of whether the photographs were taken from high in the mountains or down in the valleys. I looked past the coils of wire, the set faces of the men, the guns and horses, and locked onto something astonishing. Something that wasn't there.

I stepped out into the sunlight, scarcely able to believe what I had seen – or what I had not. I stared at the hills around me, contrasting them with the photographs. Some of the pictures had been taken here or in other places we knew, including the section of the Soˇca valley in which we were staying. Yet, where dense forests now grew, forming a high, closed canopy – in the valleys, over the hills and up the mountain walls until they shrank, many thousands of feet above sea level, into a low scrub of pines, which diminished further to a natural treeline – there had been almost nothing. The land in the photographs, taken on the western side of Slovenia during the First World War, was almost treeless.

When I say that a country is the size of Wales, I do not expect you to take that statement seriously. Wales is used as a comparison so often that it has almost become a unit of measurement. How many times have you read that 'an area of rainforest the size of Wales has been destroyed in the Amazon this year' or 'the floods have drowned a region the size of Wales', or 'the rescue services must search an area of bush the size of Wales'? But in this rare case, the comparison is not a loose one: they are almost identical in size.* Slovenia's population (2 million) is slightly smaller than that of Wales (3 million) and its gross domestic product, during the year before our visit, a fraction higher.† There the resemblances end.

---

* Slovenia, at 20,273 square kilometres, is 98 per cent of the area of Wales, which covers 20,779 square kilometres.
† Slovenia, €18,000 per head in 2009.[1] Wales: £14,800 – equivalent at the time of writing to €17,000.[2]

While the uplands of Wales have been progressively deforested over the past century, the vegetation of the hills and mountains of Slovenia has shifted in the same period from grassland and scrub to deep forest. So tall and impressive are the trees and so thickly do they now cover the hills that when you see the old wartime photos – taken, in ecological terms, such a short time ago – it is almost impossible to believe that you are looking at the same place. I have become so used to seeing the progress of destruction that scanning those photographs felt like watching a film played backwards.

We slid the raft down the bank, into the shallow water beneath an overhanging beech tree. The ripples it made rocked across the smooth water, furling up then laying out the early autumn colours – green, ginger, yellow, blue – like a roll of psychedelic linoleum. We slipped into the boat, paddled out into the middle of the river then stowed the oars. As soon as the raft felt the current it began to turn, like a fallen leaf, and to drift down the river. Neither of us said a word.

On the left, Slovenia glided past us; on the right, Croatia. Both were cloaked in deep forest. Beech, maple and aspen overhung the water and trailed their twigs in the current. On the steep limestone hills on either side of the River Kolpa, silver firs broke through the canopy of deciduous trees. Birdsong poured from the woods and rolled across the water. Otherwise, but for an occasional car passing along the narrow road on the Slovenian side and the distant rumble of a weir, there was no sound.

I lay back in the boat. The river and the sky were fringed by leaves. Around the sallows beside the water, redstarts and wagtails flickered through the mottled sunlight. A thrush passed across the river of sky above us, its wings a silver gauze against the light.

Soon the current picked up, and the first weir came into sight. To give ourselves time to inspect it before we went over, we pushed the boat onto a gravel spit.

Though it could not be so, it looked as if no human had ever trodden there. On the upstream end of the spit the smell of peppermint was so strong that I fancied I could almost see the trails of scent hanging above the bushes. It formed a hedge, waist high, that released a cloud of insects as I brushed through it. The far end of the spit, which had built up against the weir, was covered by a thicket of willow. Pushing through, I found a disused duck's nest. Warblers flitted among the branches. I struggled across to the far side, where woody nightshade hung over a derelict mill stream. Yellow stamens protruded from the dark flowers like stings. In the stream, brown trout with red and black stipples rose to kiss the surface. I watched them for a while, then pushed back through the withies to the other side of the bar, where we stared at the water sliding slickly over the lip of the rocks, before exploding into feathers of spray.

Above the weir the water looked stretched, its polished surface scarred by turbulence. More trout hung beneath it, resting their tails on the rebounding water above the rocks, eyeing the caddis flies that struggled to break free from the surface, rising and snatching them with a white flash of the mouth. The dents they made on the surface smeared over the sill.

Hearing the water crepitate along the gravel bar, watching the autumn leaves slide down towards the weir and the white water crashing over it, I thought of the reindeer carving that I loved in the British Museum. A stag and hind are struggling south across a rushing river, following the autumn herds migrating to their winter pastures. The stag has propped his chin on the hind's rump as he paddles, nostrils flared, antlers thrown back, eyes popping with effort and arousal. You can almost hear the reindeer snorting and panting, see the water lapping round their chins, dragging down their long winter coats. All this is rendered in a piece of mammoth ivory the size of a carrot, carved with a chip of flint 13,000 years ago.

We negotiated the weir in a fashion that I would struggle to describe as graceful: backwards, in a tangle of limbs and paddles. The judges who reside in my head held up their zeros.

Then we swung the boat round, and drifted through a wide, shallow stretch. Far ahead of us, someone poled across the river from Slovenia to Croatia in a punt, moving into then out of the narrow band of sunlight. We passed her house. Overhanging the water was an apple tree. I could see the red and green apples, turning slowly in the eddies along the bank, occasionally flaring in the light half a mile downstream. I scooped a few out of the river and we ate them as we lay in the boat.

After a few more weirs, which we crossed with a little more dignity, we drifted into a deep, narrow chasm, between limestone bluffs. I stared down into the water. Though it was some three fathoms deep here, the river was so clear that I could see the bottom, and the shadows of the fish which passed over it like unformed thoughts.

As we emerged from the gorge I noticed a creature unlike anything I had seen before. Sickly grey with large black spots, a big head with a hooked jaw, the cold yellow eyes of a wolf, as long and lean as a pike, it continued, unafraid of us, to patrol the bank, hunting. It was a *huchen*, the predatory landlocked salmon of the Danube catchment. At three or four pounds this one was still an infant; some reach sixty.

The rivers further north, which drain into the Adriatic, also contain monsters. The marbled trout which inhabit them, like the *huchen*, grow to sixty pounds. A fisherman I spoke to on the banks of the River Soˇca told me that sometimes when he had hooked a grayling and was bringing it to the net, a monstrous trout would loom out from behind a boulder, snatch it off the hook and swallow it whole. As the forests of Slovenia had recovered, so had the rivers. The soil was bound up by the roots of the trees and could no longer be stripped from the land, so they now ran clear. They were contaminated by neither pesticides nor fertilizers, and, because the woods slowly released the water that fell on them, they did not suffer the worst extremes of flood or drought.

Tomaž Hartmann drove for almost an hour along a forest track through Kočevski Rog. The woods of beech and silver fir towered over us, in places almost touching across the road. Their roots sprawled over mossy boulders. They rolled down into limestone sink-holes: karstic craters. Karst topography – weathered limestone landscapes of chasms and caves, sinkholes, shafts and pavements – is named after this region of Slovenia, which is sometimes called the Kras or Karst plateau. The word means barren land. When Karst landscapes are grazed they are rapidly denuded, but it was hard to connect the term with what I now saw.

Where the road clung to the edge of a hill, I could see for many miles across the Dinaric Mountains. The view was framed by the tops of the trees beneath us, through which the sunlight filtered. The mountains rambled across the former Yugoslavia, fading into ever fainter susurrations of blue. The entire range was furred with forest. Where the road sank into a pass, the darkness closed around us. Through the trunks I could see the air thicken, shade upon shade of green. A few yards from the road a fox sat watching us. Its copper fur glowed like a cinder in the shadows, which cooled to charcoal in the tips of its ears. It raised its black stockings and loped away into the depths. Woodpeckers swung along the track ahead of us.

The leaves of the beeches glittered in the silver light above our heads. The great firs grazed the sun, straight as lances. They looked as if they had been there for ever.

'All this,' Tomaž told us, 'has grown since the 1930s.'

He parked the car and we set off up a forest trail. Mushrooms nosed through the leaf litter beside the path. Saffron milk caps, orange and sickly green, curled up at the edges like Japanese ceramics. Dryad's saddle, sulphur tuft and cauliflower fungus accreted around rotting stumps. Russulas – scarlet, mauve and gold – brightened the forest floor.

Tomaž led us up a tumbled limestone slope towards a stand of virgin forest, the ancient core of the great woods which had regenerated over the past century. As we climbed, we stepped into a ragged fringe of cloud. Sounds were muffled. The trees loomed darkly out of the fog. Tomaž spoke as we walked about the dynamism of the forest system: how it never reached a point of stasis, but tumbled through a constant cycle of change. He had noticed some major shifts, and knew that, as the climate warmed, there would be plenty more. Though he described himself as both a forester and a conservationist, he had no wish to interrupt this cycle, or to seek to select and freeze a particular phase in the succession from one state to another. He sought only to protect the forests, as far as his job permitted, from destruction by people.

Now in his sixties, he had worked in these forests for most of his adult life. He was a gentle, engaging man, with a mild face and a white beard, who appeared to be at peace with his life. Working in the forests, he said, had, with his family, given him all the delight and purpose in life a man could wish for. When he was not working, he made ephemeral sculptures in the woods, from leaves and snow and fallen branches.

Ahead of us something dark and compact shot across the path in a blur and disappeared into the undergrowth: probably, Tomaž said, a young wild boar. Then, though it was not clear where the transition occurred, we found ourselves in the primeval core of the forest. The trees we had walked past until then were impressive, but these were built on a different scale. The beeches grew, unbranched for one hundred feet – smooth pillars wrapped in elephant skin – until they blossomed, like giant gardenias, into a leafy plateau in the forest canopy. Silver firs pushed past them, the biggest topping out at almost 150 feet. Only where they had fallen could you appreciate the scale of their trunks.

The forest had entered a cycle Tomaž had not seen before, in which many of the giants had perished. Some had died where they stood, and remained upright, reamed with beetle and woodpecker holes, sprouting hoof fungi and razorstrops. They looked as if a whisper of wind could blow them down. Others now stretched across the rocks and craters, sometimes blocking our path, sometimes suspended above our heads. Among the trunks lying on the ground, some were so thick that I could scarcely see over them. Where they had fallen, thickets of saplings crowded into the light. Seeing the profusion of fungus and insect life the dead wood harboured, I was reminded of the old ecologists' aphorism: there is more life in dead trees than there is in living trees. The tidy-minded forestry so many nations practise deprives many species of their habitats.

On a large rotten log which had lost its bark and was now furry with green algae, Tomaž showed us two sets of four white marks: deep parallel scratches where a bear had sharpened its claws. He told us that he had seen plenty of bears in the forest, but – though they are abundant here – never a wolf or a lynx. Just knowing that they were there enriched and electrified every moment he spent in the forest. I felt it too, like a third beat of the heart. The forest seemed to bristle with possibility. Here, to mangle Auden, nature's jungle growths were unabated, her exorbitant monsters unabashed.[3] This great rewilding, Tomaž explained, was the accidental result of a series of hideous human tragedies.

Some 150 years ago, just 30 per cent of the Kočevje region, 95 per cent of which is now forested, was covered by trees. Much of the forest was preserved by the Princes of Auersperg as hunting estates. So obsessed by hunting were they, as princes often seem to be, that they and the other great lords of the Habsburg monarchy in Slovenia and Croatia drew up an official declaration of friendship with the bear, signed and stamped with their great seals, in which they agreed to sustain its numbers so that they could continue to pursue it. The role the bears played in this negotiation is unrecorded.

The revolutions of 1848 brought feudalism to an end in central Europe. Local farmers lost their rights to graze common land, but acquired their own private plots. At around the same time, imports of cheap wool from New Zealand began undermining the European industry. By the end of the nineteenth century, many peasant farmers had sold their land and either moved to the cities or emigrated to America. The Depression of the 1930s further extended the woods – to around 50 per cent of Kočevje – as more people departed. But the greatest expansion of the forest took place as a result of what happened in the following decade.

Most of the population of south-western Slovenia – around 33,000 people – was ethnic German. They kept sheep and goats in the hills and ran much of the trade in the towns. Under King Alexsander's autocracy in the ten years before the Second World War, the Germans of Yugoslavia, around half a million in total, suffered discrimination and exclusion. In response, many of them joined German nationalist movements, some of which soon allied themselves to the Nazis. By 1941, when Hitler's army suddenly invaded Yugoslavia, over 60 per cent of its ethnic Germans had joined an organization, the Kulturbund, which became absorbed into Himmler's euphemistically titled Volksdeutsche Mittelstelle, or Ethnic Germans' Welfare Office.*

Hitler ceded south-western Slovenia to Italy and the Nazis forcibly relocated many of the Yugoslav Germans to the Third Reich, to preserve their 'ethnic purity' and protect them from attacks by partisans. Some of the Germans of Kočevje were transferred to eastern Slovenia, some removed to other lands under German rule.

The horrors of the 1990s in Yugoslavia were a faint echo of what happened there during the Second World War. Many ethnic and religious groups committed atrocities, conducting expulsions, massacres and genocidal cleansing which stand out even among the other disas-

---

* I have expanded on the account provided by Tomaž and other Slovenians I spoke to, drawing in particular on materials published by the Institute for Research of Expelled Germans.[4]

ters of war. Almost a million people died in the Yugoslavian civil strife triggered by the Nazi invasion. Some of these great crimes were committed by the Prinz Eugen Division of the SS, among whose members were Yugoslavian ethnic Germans. They massacred Jews, partisans and communists and people believed to sympathize with them.

After the Axis forces were routed, Marshall Tito's communist government found it convenient to blame ethnic Germans for many of the horrors perpetrated by other people. This was, it seems, easier than facing the truth: that atrocities were committed by Croats, Serbs, Bosnians, Albanians, Hungarians, Nazis, communists, monarchists, Orthodox Christians, Catholics and Muslims. Almost all the Yugoslavian Germans who did not flee the country with the Axis armies were either expelled by Tito's government or interned, often in forced labour camps. Some were taken by the Soviet Union's Red Army to camps in the Ukraine. Within a few years of the end of the war in Yugoslavia, the German population had dropped by some 98 per cent.[5]

Many others who collaborated with the Third Reich were killed. The six battalions of the Slovenian Home Guard fled with the retreating German troops to Austria in May 1945.[6] They were forcibly repatriated by the British. Driving with Tomaž through the forests of Kočevski Rog, I had seen beside the road great trunks like totem poles carved by the sculptor Stare Jarm into the tortured figures of Christian martyrs. They marked the sinkholes beside which some thousands of collaborators were lined up and machine-gunned. The partisans then used explosives to make the craters collapse, burying the corpses.

The barren lands of Kočevje, whose population had been relocated and dispersed first by the Nazis then by the socialist government and the Red Army, were never recolonized. When the farms were abandoned and the pastures no longer grazed by sheep and goats, the seed which rained into them from the neighbouring woods was allowed to sprout once more. The land has been repopulated by trees.

In the Soča valley, in north-western Slovenia, Jernej Stritih, a clever, laconic head of department in the Slovenian government, with a thick beard and splendid moustaches, whom we had befriended in Ljubljana, took us to a restaurant a friend of his ran in the front room of his farmhouse. The proprietor owned a small herd of sheep, which were kept for show and to make cheese to sell to tourists. We had seen them on display that morning in the Trenta Fair, massive beasts weighed down by trailing yellow coats. They had won first prize, and now a large gilt cup stood on a table, glimmering in the low brown light, while he, in a leather waistcoat and bushy side-whiskers, drank and talked with his friends. From time to time he would stop talking and, almost as if he were unaware that he was doing so, bend down to play the dulcimer on the table before him, while the other men continued their conversation.

As we ate, Jernej explained that our host was one of the last shepherds in the region. Because there was no longer any arable production in the valley, the few remaining sheep could stay in the lowlands and were never led into the mountains. Here, by contrast to Kočevje, there had been no mass dispossession of local people. A different social tragedy had been engineered. In the 1950s, he told us, Tito had banned the goat. The ostensible purpose was to protect the environment, but doubtless he also sought to drag the peasantry out of what Marx and Engels called its 'rural idiocy' and press it into the urban proletariat. (The peasants of eastern Europe had perversely failed to fulfil the *Communist Manifesto*'s prediction that they would 'decay and finally disappear in the face of modern industry'.) Without goats, which browsed back the scrub, the pastures became unsuitable for sheep.

The rewilding of the western side of Slovenia, the rapid regrowth of forests there and the recovery of its populations of bears, wolves, lynx, wild boar, ibex, martens, giant owls and other remarkable creatures, took place at the expense of its human population. This is not to suggest that it continues to generate social tragedy. On the contrary, this region has become a lucrative destination for high-end tourism, which supports what was, when we visited, a buoyant local economy. Slovenia's rivers are said to offer the best fly-fishing in Europe. I spent a day working my way up a few miles of the Soča, a glorious tumble of turquoise water winding through limestone gorges, watching a tiny dry fly bouncing down the glides and eddies. To get back to where I had begun, I hitched a lift along the valley road. I was picked up by a local van driver.

'You're fishing, when the water's so high?'

'It's the only chance I have.'

'It's unfishable today. How did you do?'

'I caught ten.'

'Just as I said. Unfishable.'

The forests and their wildlife, the mountains, repopulated by ibex and chamois, the caves with their endemic species of blind salamander, known to locals as the human fish on account of its smooth pink skin, the rivers with their steady flow and excellent whitewater rafting, the extraordinary beauty of this regenerated land, draw people from the rest of Slovenia, from all over Europe and beyond. As I talked to many Slovenians, it became clear that the integrity of the natural environment was now a source of national pride.

The forests give rise to other industries too. We happened to pass through Ribnica, on the way to Koˇcevje, on the day of the annual wood market. We stopped for a few hours and walked among perhaps a hundred stalls, selling snaths and grass rakes, scratters and presses, besoms and brooms, trugs and baskets, stools and barrels, cradles and rocking horses, racks and rolling pins. Men sporting waistcoats, sugar-loaf hats and enormous moustaches eased their way through the crowds, playing their accordions. The market square had been set with tables, and we joined a municipal barbecue that fed hundreds. The woodwares being sold that day, we were told, were a small part of the output of a thriving cottage industry begun in the Middle Ages, when the Habsburg emperor granted the region's population unlimited rights to sell its wares throughout the empire, in the hope of alleviating local poverty. No one would become a millionaire this way, but it kept people and their communities alive.

None of this is to deny a disquieting truth, however. Slovenia is just one example of a global phenomenon. Most of the rewilding that has taken place on earth so far has happened as a result of humanitarian disasters.

Throughout the Americas – North, Meso and South – the first Europeans to arrive in the sixteenth century reported dense settlement and large-scale farming. Some of them were simply not believed. Francisco de Orellana and Brother Gaspar de Carvajal, who travelled the length of the River Amazon in 1542, claimed that they had seen walled cities in which many thousands of people lived, raised highways and extensive farming along its banks.[7] When later expeditions visited the river they found no trace of them, just dense forest to the water's edge and small scattered bands of hunter-gatherers. Orellana and Carvajal's reports were dismissed as the ravings of fantasists, seeking to boost commercial interest in the lands they had explored.

It was not until the late twentieth century that investigations by archaeologists such as Anna Roosevelt[8] and Michael Heckenberger[9] suggested that his accounts were probably accurate. In parts of the Amazon previously believed to have been scarcely habited Heckenberger and his colleagues have found evidence of garden cities surrounded by major earthworks and wooden palisades, built on grids and transected by broad avenues. In some places they have unearthed causeways, bridges and canals. The towns were connected to their satellite villages by road networks which were planned and extensive. These were advanced agricultural civilizations, maintaining fish farms as well as arable fields and orchards.[10] It appears that European diseases – smallpox, measles, diphtheria, the common cold – brought to the Caribbean coast of South America by explorers and early colonists, passed down indigenous trade routes into the heart of the continent, where they raged through densely peopled settlements before any other Europeans reached them. So ferocious is the vegetation of the Amazon that it would have obliterated all visible traces of the civilizations its people built within a few years of their dissolution. The great *várzea* (floodplain) forests, whose monstrous trees inspired such wonder among eighteenth- and nineteenth-century expeditions, were probably not the primordial ecosystems the explorers imagined them to be.

The same goes for the fauna and flora of the rest of the Americas. Early hunter gatherers wiped out most of the megafauna of the western hemisphere. Some Native American civilizations – such as the Maya in the Yucatán – destroyed large tracts of forest. Places which were later seen as *terra nullius* or *informem terris*,* virgin lands unshaped by man, turn out to have been densely populated before all but the very first explorers arrived. As the writer Ran Prieur observes in the journal *Dark Mountain*:

---

* *Terra nullius*, a concept formalized in Roman law, means land belonging to no one. *Informem terris*, a phrase that might have been coined by Tacitus, means shapeless or dismal lands.

The incredible biological abundance of North America was also a post-crash phenomenon. We've heard about the flocks of passenger pigeons darkening the sky for days, the tens of millions of bison trampling the great plains, the rivers so thick with spawning salmon that you could barely row a boat, the seashores teeming with life, the deep forests on which a squirrel could go from the Atlantic to the Mississippi without touching the ground. We don't know what North America would have looked like with no humans at all, but we do know it didn't look like that under the 'Indians'. Bone excavations show that passenger pigeons were not even common in the 1400s. 'Indians' specifically targeted pregnant deer and wild turkeys before they laid eggs, to eliminate competition for maize and tree nuts. They routinely burned forests to keep them convenient for human use. And they kept salmon and shellfish populations down by eating them, and thereby suppressed populations of other creatures that ate them. When human populations crashed, nonhuman populations exploded.[11]

Gruesome events – some accidental, others deliberately genocidal – wiped out the great majority of the hemisphere's people and the rich and remarkable societies they created. In many parts of the Americas the only humans who remained were – like the survivors in a post-holocaust novel – hunter-gatherers. Some belonged to tribes which had long practised that art, others were forced to reacquire lost skills as a result of civilizational collapse. Disease made cities lethal: only dispersed populations had a chance of avoiding epidemics. Dispersal into small bands of hunter-gatherers made economic complexity impossible. The forests blotted out memories of what had gone before. Humanity's loss was nature's gain.

The impacts of the American genocides might have been felt throughout the northern hemisphere. Richard Nevle and Dennis Bird at Stanford University have speculated that the recovering forests drew so much carbon dioxide out of the atmosphere – about ten parts per million – that they could have helped to trigger the cooling between the sixteenth and seventeenth centuries known as the Little Ice Age.[12] The short summers and long cold winters, the ice fairs on the Thames and the deep cold depicted by Pieter Brueghel might have been caused partly as a result of the extermination of the Native

Americans. (There is little danger that rewilding would cause a little ice age today: human activity has raised carbon dioxide concentrations in the atmosphere by over one hundred parts per million.)

If another fascinating speculation is correct, Native American civilization may have begun with a similar impact. The biologist Felisa Smith proposes that the extermination of the American megafauna by Mesolithic hunters was responsible for another mini ice age, the Younger Dryas,* which began 12,800 years ago and lasted for 1,300 years.[13]

The wild herbivores of the Americas were, like cattle and sheep, magnificently flatulent. Smith calculates that they produced around 10 million tonnes of methane a year. Methane is a greenhouse gas, active for a shorter period than carbon dioxide, but, while it persists, around twenty times as powerful. The sharp decline in methane production when the large herbivores became extinct might have been sufficient to account for the collapse in temperatures (a global decline of between 9 and 12° Centigrade) at about the same time. If this is correct (it is one of a number of competing explanations), the history of the first peoples of the Americas was bookended by catastrophe and climate change.

In his masterpiece *Landscape and Memory*, Simon Schama explores the narratives and impulses which gave rise to what could be described as Nazi rewilding projects.[14] One of the most powerful myths of German nationhood arose from a remarkable event that took place 2,000 years ago in the great primeval forests around the River Weser, that the Germans later called the Teutoburger Wald. The people of these forests, according to the Roman historian Tacitus, were wild and free. They worshipped beneath the trees and offered human sacrifices to the god of the woods. Uncorrupted by luxury, dressed only in pelts and cloaks, they were, he claimed, chaste, tough and massive. These Cheruscan tribesmen were the people organized by the man Tacitus called Arminius and the Germans call Hermann.

Hermann was the son of a German chief captured by the Romans. He was recruited into the Roman army and rose through its ranks,

* The name refers to a tundra flower, *Dryas octopetala*, that became common in this period.

but he never forgot his tribal identity. He raised a rebellion in the *urwald*, and in AD 9 the wild men he commanded ambushed the Roman army commanded by Publius Quintilius Varus, which was marching through the great forest to its winter quarters. The Cheruscan savages trapped Varus's 25,000 men between swamps and wildwood, and speared to death all but a handful of the decadent, complacent empire's troops. From this unlikely victory a compelling but ultimately lethal myth was born.

From the late fifteenth century onwards, Germans began to portray themselves as the descendants of wild and natural beings who pursued an uncorrupted existence in a woodland arcadia. By the mid-eighteenth century, the forests in which Hermann defeated the civilized Romans began to embody the authentic fatherland – raw, free and strong.

Wald and *Volk* – forest and people – were explicitly connected by Nazi ideologists. In 1941, when the German army launched its attack on the Soviet Union and overran eastern Poland, Hermann Göring, commander-in-chief of the Luftwaffe, seized the Białowieża Forest – the *urwald* preserved through the centuries as a royal hunting estate – and declared it his private property. The government conservation department he had established then set to work to create a vast national park around the ancient forest, from which the people were cleared (and many murdered) with customary Nazi cruelty.[15] The land was rewilded by brute force.

Göring's brutalities in eastern Poland were an extreme form of what the Normans did in England. Their forest law annexed large tracts of countryside. 'Forest' meant not a place where trees grew but a place *foris* – or outside – the usual rule of law. Elsewhere the use of the land was often widely shared, but these tracts (some of which were treeless) were subject to the harsher and less accommodating demands of the royal hunt. In some cases forest law cleared the inhabitants out, in others it curtailed their rights and reduced their living. Like Göring, William I and his court were obsessed by the chase, and they saw the capture and creation of new hunting grounds as one of the perquisites of conquest. The forest laws were brutally extended by the Black Acts of the eighteenth century, documented in E. P. Thompson's book *Whigs and Hunters*.[16] The new hanging offences they created were designed to discourage local people from defending themselves against the creeping encroachment upon their crops and rights by the king's deer and the royal hunt.

The principles of forest law were exported to the British colonies. In Kenya, the colonial authorities evicted local inhabitants from land they designated as game reserves, which later became national parks and nature reserves. The evictions were justified on the grounds that the presence of people and their domestic animals was incompatible with the preservation of wildlife. This, coming from a settler population which had made an abattoir of the savannahs, was rich. In fact it was only because the indigenous people had not destroyed the herds of wild animals with which they had lived up to and beyond the arrival of the British in East Africa that the Europeans wanted to annex and conserve their lands. Only wardens, rangers and paying tourists were allowed into the parks and reserves. If the people who had lived on those lands tried to return to them, they would be treated as trespassers or poachers.

When I worked in East Africa in the early 1990s, this process of enclosure was being extended in both Kenya and Tanzania. Already the Maasai had lost all but two of their dry season grazing lands, and were now in danger of losing the remainder. With the help of a British conservation group, the Maasai had just been expelled from the Mkomazi Game Reserve in northern Tanzania. They were dumped in the surrounding farmland, where they were promptly arrested for criminal trespass and fined. They tried to return to the reserve, but when they arrived they were once more arrested for criminal trespass and fined. Their cattle died of starvation.

In Kenya I met Maasai herdsmen who had been hospitalized by rangers working for the Kenya Wildlife Service when they tried to return to their dry-season pastures. When I challenged the then-director of the service, Dr Richard Leakey, about these policies, he produced a brutally utilitarian defence of enclosure and clearance. 'The setting aside of land for the purpose of wildlife conservation, to support the tourist industry, is a strategic issue. The morality of evicting people from land, whether it's to establish a wheat scheme, a barley scheme, hydroelectric scheme or a wildlife tourist scheme is the same. Basically nation states have got to function.'[17]

The campaign to create the Yellowstone National Park in the United States – the world's first national park – was also assisted by potential revenues from tourism. Though Yellowstone's champions, such as Thomas Meagher, Cornelius Hedges and Ferdinand Hayden, were motivated by their love for the land, the proposal was to a large extent driven and financed by Jay Cooke, owner of the Northern Pacific Railroad.[18] He hoped that the tourist trade would boost his railway's income. (Cooke failed to benefit from its establishment in 1872, however, as his company collapsed in 1873.)

The act which created the park states that the land

> is hereby reserved and withdrawn from settlement, occupancy, or sale under the laws of the United States . . . and all persons who shall locate, or settle upon, or occupy the same or any part therof, except as hereinafter provided, shall be considered trespassers and removed therefrom.[19]

The provision was necessary to preserve the character of the land from encroachment by European Americans, at a time in which the West was being rapidly transformed. But Congress overlooked the fact that it had been settled for some 11,000 years, and appears still to have been used by the Crow, the Shoshone Tukadika and the Blackfoot.* They too were transformed into trespassers in the park and were eventually removed therefrom. The act preserving Yellowstone – and its clearance of native people – became the model for the creation of national parks throughout the Union, and in many other parts of the world.

Though the mores of modern wildlife agencies are not comparable to those of the Nazis, there are common themes, which long predate the Third Reich and which have continued long beyond its collapse, informing a process that could be described as forced rewilding.

Since Schama's book was published, further research has cast new light on Nazi attitudes to nature and attempts at rewilding. Fascinating

---

* Not everyone accepts this account. Susan Hughes claims that 'the Sheepeaters [Shosone Tukadika] as depicted in northwestern Wyoming folklore are predominantly a myth derived from the medieval wild man and an Indian stereotype passed down through colonial history . . . a permanent band of Sheepeaters in Yellowstone National Park may never have existed.'[20]

papers by Boria Sax and Martin Brüne summarize recent discoveries about the dark side of Professor Konrad Lorenz.[21] Lorenz, an Austrian, is widely considered to be the founder of the modern science of animal behaviour (ethology). His work in this field won a Nobel prize. But we now know that he was also responsible for helping to formulate some of the unscientific tenets of Nazi ideology. He advocated a programme of eugenics whose purpose was to rewild human nature, by stripping people of what he considered to be the genetic legacy of civilization.

Lorenz sought scientific justifications for Friedrich Nietzsche's attempt to equate the civilization of humans with the domestication of animals. In both cases, Lorenz claimed, the result was genetic decline and the disruption of what Nietzsche celebrated as instinctive behaviour, leading to social breakdown, degeneracy, indiscriminate breeding, a lack of patriotic enthusiasm and eventual human extinction. He appeared to endorse the view of the ancient Greeks that, as he put it, 'a handsome man can never be bad and an ugly man can never be good'.[22] He listed the physical characteristics which he said were caused by both human civilization and the domestication of animals – rounded heads, shortened limbs, pot bellies – which happened to correspond with popular Nazi stereotypes of Jewish physiognomy. He coined a term for this supposed transformation: *Verhausschweinung*, or pig domestication.

Immediately after the *Anschluss* (the German annexation of Austria) in 1938, Lorenz joined the Nazi party. He became a member of its Office for Race Policy and proposed a programme of eugenics which exceeded even the scheme overseen by Heinrich Himmler. Lorenz believed that humans could be bred to meet not only a physical ideal but also an ethical one. He argued that it was not just those with 'domesticated' physiques who should not be allowed to reproduce, but also those possessed of 'domesticated' instincts. Those selected for breeding, on the other hand, would form not just a master race but a master *species* of instinctive, wild beings. He advocated the 'extermination of ethically inferior people' and conducted a study of the children of marriages between Germans and Poles, which led to those assessed as genetically deficient being dispatched to concentration camps.[23]

His notions of racial purity corresponded to Nazi conceptions of wildness. By sharp contrast to most European thinking in the nineteenth century, Sax explains, the Nazis saw nature not as lawless and chaotic but as ordered and standardized. They compared themselves to wild predators which, they believed, had an inherent right to rule the ecosystem. After the war, Lorenz pursued this analogy, though now in coded form. He claimed, wrongly, that domestic dogs had two genetic origins: the northern wolf and the Mesopotamian jackal. Dogs descended from wolves, he believed, inherited the characteristics of animals which form 'a sworn and very exclusive band which sticks together through thick and thin and whose members will defend each other to the very death'.[24] Dogs descended from jackals, by contrast, were obedient, but infantile and lacking in loyalty. These traits corresponded to Nazi characterizations of the 'Aryan' tribes of the North from which they claimed the Germans were descended, versus the 'degenerate' peoples of the South among whom, they maintained, the Jews arose.

An attraction to large predators often seems to be associated with misanthropy, racism and the far right. The extract from D. H. Lawrence's poem *The Mountain Lion* with which I began this chapter hints at this conjunction of interests. In his book *The English Novel*, Terry Eagleton notes that while Lawrence 'regarded fascism as a spurious solution to the crisis of middle-class civilization', there are elements of his thinking – racism and anti-Semitism among them –

> which sail perilously close to the fascist creed. . . . at his most danger-
> ous he invites us to discard rationality as itself a kind of alienation,
> and think with the blood and racial instincts instead. It was this aspect
> of his work which Bertrand Russell considered led straight to
> Auschwitz.[25]*

The British millionaire John Aspinall, who died in 2000, made his money running gambling dens. He made his name spending this money on the zoos he founded – Howlett's and Port Lympne in Kent – where his breeding programmes enjoyed great success. He fetishized the tigers he kept. He encouraged his keepers to interact freely with

---

* I would question the idea that Lawrence invites us to discard rationality, on the grounds that we do not have a great deal to discard (as the work of researchers such as Jonathan Haidt and Antonio Damasio shows). What he invites us to discard, and what I think Eagleton and Bertrand Russell are talking about, is universalism. If blood and culture are allowed to outweigh the consistent application of universalist principles (in particular the golden rule), this can become a licence to trample on other people.

them, with the result that three of them were mauled to death (two others were trampled to death by his elephants). When he was dying of cancer, Aspinall tried to induce his tigers to kill him too.

He believed that the human race was 'vermin'[26] and announced that 'Britain's population problem can be solved by beneficial genocide'.[27] He maintained that 'the concept of the sanctity of human life is the most damaging that philosophy has ever propagated'[28] (his zoo-keeping policy was, it seems, consistent with that belief). He professed himself a supporter of Hitler's views on eugenics,[29] and described his third wife as 'a perfect example of the primate female, ready to serve the dominant male and make his life agreeable'.[30] He worked with Mangosuthu Buthelezi to undermine the African National Congress and forestall majority rule in South Africa. With Lord Lucan (who later disappeared after allegedly bludgeoning his children's nanny to death) and the financier Sir James Goldsmith, he discussed the possibility of launching a military coup against Harold Wilson's Labour government.[31]

Joy Adamson's book *Born Free*, published in 1960, about bringing up and then releasing a pet lioness in Kenya, was wildly successful. The portrayal of her character and behaviour in the book and the Oscar-winning film that dramatized it was pure fiction. In reality she possessed a strong suite of what might have been psychopathic traits. She got what she wanted through a combination of manipulation and volcanic eruptions of temper. While devoting great care and attention to the lions, leopards and cheetahs she looked after, she appeared to have few scruples about the way she treated people, especially her African servants, and little understanding of the hurt she caused.

Her biographer, Caroline Cass, records that when a boy who worked in her kitchen took too long to deliver her tea, Adamson threw it in his face, scalding him.[32] When her cook spoilt the soup, she dragged him before the magistrate, and demanded, unsuccessfully, that he be beaten by the police. She forced another servant, who was seriously injured with third-degree burns as a result of an accident, to walk eight miles to the clinic for treatment, refusing to drive him as a punishment for his carelessness.

Adamson's first public lectures on the world tour she began in 1961 were entitled 'Man: the inferior species'. She threatened to shoot the keepers in Sydney zoo, after alleging that they were mistreating their lions. In Kenya she demanded that the colonial authorities give her 30,000 acres of land belonging to native people so that her pets could use it. When she was eventually murdered by a former servant, the investigation was delayed by a surfeit of possible culprits. Cass notes that 'Few people were surprised that Joy may have been killed by an African. The general opinion was that Joy got what she deserved, treating them so appallingly, forgetting to pay their wages and dismissing them with extreme rudeness and little regard for their welfare.'[33]

The Nazis' interest in re-creating what they considered to be the natural order was not confined to predators. They wanted to restore the entire ecology of the primeval forests. A reinvented *urwald*, they believed, required an *urox*.

The last giant aurochs died in Poland in 1627. The date is recent enough for the animal still to haunt Polish culture and language. Men of impressive physique, for example, are not 'built like a brick shithouse', as they are in Britain, but 'built like an aurochs'. It is a good simile.

A quarter of a century ago I was taken by the archaeologists who had just discovered it to a swallowhole in the Mendip Hills which had been used by Bronze Age people as a rubbish dump. Above the ground, the hole was almost invisible, a crack in the rocks screened by bracken and brambles. I squirmed backwards into the cleft. My feet found the wire ladder the archaeologists had hung from the lip. When I reached the bottom and planted my boots among the limestone boulders, I turned and scanned the chamber with my head torch.

The cavern was high enough to stand in. The walls and floor and everything that lay on it were encrusted with calcite crystals that glittered in the torchlight. Beneath the mineral frost I could make out shapes in the heap of treasure spilling down the ground that sloped away into the darkness: broken pots, skulls, bones of many shapes and sizes. The air was cool and damp, but not musty. It smelt only of rock and water.

One of the archaeologists bent down and picked something up. He passed it to me. 'What's this?' he asked. It was a flattish, winged bone, about the length of my palm, pierced by a large hole.

'Atlas vertebra.'

'Of course. But what of?'

'Er, red deer?'

'No, it's a Bronze Age cow. Their cows were smaller than they are today – about the size of Dexters. Now what's this?'

He lifted it up and I took it with both hands. It must have been eight inches across and have weighed a couple of pounds. I stared dumbly at it in the light of my torch.

'Atlas vertebra of, of – a mammoth?'

'What, in the Bronze Age?!'

'I – I haven't the faintest idea.'

'It's the same species as the first one.'

He told me that this was the animal from which domestic cattle were first bred. The wild cows were slightly bigger than those of modern cattle, but the bulls were massively greater: vast, heavy-shouldered animals with monstrous horns. As I turned the bone over in my hands, feeling its weight, feeling the years fall away, feeling myself, in that cave of Bronze Age junk, fall with them, I experienced what seemed like an electric jolt. The great weight of the bone, the knowledge of what it was, the sense – so clean and new it seemed – that the beast whose head it bore might have been hunted and slaughtered not 3,000 years ago but so recently that I could almost reach out and place my palm on the sweat and hair of its cooling flank ran through my arms and fulminated in my head, almost with a flash of light. It might have been at this point that the imaginative journey began which, many years later, led to this book.

The brothers Ludwig and Heinz Heck, respectively the directors of the Berlin and Munich zoos during the period of Nazi rule – a time in which zookeeping was a political activity – were not content with reconstructing the aurochs in their minds. They wanted to create a real one.[34] Like the scientists in *Jurassic Park*, they sought to resurrect this animal, as well as the ancestral horse, from genetic material; but in this case the material carried by the wild animal's descendants. As Konrad Lorenz hoped to do with human beings, they tried to strip cattle of their domestic traits so that the purified, uncivilized beast within could break, pawing and roaring, out of its degenerate husk.

As was so often the case with the declarations the Nazis made, the success the Heck brothers claimed for their attempts at genetic reversal was exaggerated. They maintained that, in the space of just twelve years, they had re-created the aurochs. All they did, in reality, was to produce a cow whose coat roughly resembled that of its wild ancestor, but which was a good deal smaller, had different proportions and would not breed true.[35] This disappointing creature, which bore as much resemblance to the giant aurochs as Himmler did to the 'Aryan' beauty he exalted, would, they proclaimed, help to restore the true German ecosystems degraded by the assaults of civilization. Ludwig Heck released some of these ersatz aurochs into the Białowieża Forest Göring had seized.

The descendants of these animals are now being used in a rewilding project in the Netherlands, on a large polder at Oostvaardersplassen, which has none of those political connotations. There they range freely within a reserve of 5,000 hectares, without veterinary treatment, shelter or feed, reprising the role the aurochs might once have played (though with the crucial differences that its predators and some of its competitors are missing, and that it cannot migrate).[36] The man who founded the project, Frans Vera, chose Heck cattle partly for their hardiness and partly, it seems, because of the public interest their unusual appearance would generate.[37]

Simon Schama rightly warns us against making 'an obscene syllogism: to imply in any way that modern environmentalism has any kind of historical kinship with totalitarianism'. Nevertheless, the forced rewildings which have taken place elsewhere offer a pungent warning of how this project could go badly wrong if we are not mindful of its hazards and antecedents. Rewilding must not be an imposition. If it happens, it should be done with the consent and active engagement of the people who live on and benefit from the land. Governments must not create, as they have done in East Africa and Botswana, a paradise for the rich from the lands of the poor. If a rewilding scheme requires forced dispossession, it should not go ahead.

There is no need for coercion. Through the proposals I have suggested and the changes that are likely to take place anyway, in the uplands of Britain and Europe, some other parts of North America and some other regions of the world, the large-scale restoration of living systems and natural processes can take place without harming anyone's interests. This will, I believe, enhance our civilization, enrich and rewild our own lives, introduce us to wonders which, in these bleak lands, now seem scarcely imaginable.

12

The Conservation Prison

What would the world be, once bereft
Of wet and of wildness? Let them be left,
O let them be left, wildness and wet;
Long live the weeds and the wilderness yet.
                              Gerard Manley Hopkins
                                         Inversnaid

I learnt my ecology in the tropics. I studied the subject as part of my degree, then applied it, to a small extent, when I worked for a couple of years at the BBC's natural history unit. But it was not until I left my own country, first for West Papua, then Brazil, then East Africa, that I began fully to appreciate this marvellous science. Only when I lived among ecosystems which retained many of their trophic levels, their diversity and dynamism, did I begin to understand how the natural world might work.

In the Amazon I fell in with a group of scientists working at the frontiers of the discipline, and shared the excitement of some of their discoveries. Their work was beginning to transform our comprehension of the living planet. The lesson I learnt repeatedly, in all three regions, was that much of the diversity and complexity of nature could be sustained only if levels of disturbance were low. Major intrusions, such as clearing trees and raising cattle, quickly simplified the ecosystem. This seems so obvious that it should scarcely need stating.

Coming home, it took me a while to notice something odd. Here, many conservationists appear to believe the opposite: that the diversity, integrity and 'health' of the natural world depend upon human intervention, often intense intervention, which they describe as 'management' or 'stewardship'. More often than not, this involves clearing trees and using cattle and sheep to suppress the vegetation. To a lesser extent, the same belief prevails in several other parts of the rich world. Some of our conservation groups appear to be not just zoophobic but also dendrophobic: afraid of trees. They seem afraid of the disorderly, unplanned, unstructured revival of the natural world.

On a cool, blustery day in June, I travelled up the mountain road between Machynlleth and Llanidloes to visit the nature reserve that is said to exemplify the delights of the Cambrian Mountains. Glaslyn is described by the group that owns it as 'Really Wild! . . . not only is this the biggest reserve currently managed by the Montgomeryshire Wildlife Trust, but it is also the wildest and most regionally important site.'[1] I expected to find an oasis, a fecund sanctuary in the Desert. Four years living on the edge of the Cambrians had not yet taught me to curb my enthusiasm.

As I parked the car beside the road, I heard a skylark pouring its song from the sky. Clouds scudded across the sun, catching the cold north-westerly in their sails. I set off down the track towards the lake at the heart of the reserve. I could see it gleaming amid the dark heather, like the water in the bottom of an old copper bowl.

Before I reached the path that would take me down to the lake, I vaulted a fence and struck out across the heath. Nowhere was the heather more than a foot high. There were a few tufts of bog cotton, like white blusher brushes, mounds of moss and cropped bilberry, some sparse constellations of tiny bedstraw flowers, scrappy little stalks of ling – and tormentil everywhere. That, in this 'really wild' reserve, was all. I was astonished, but the clues were not hard to spot: sheep shit, all over the heath. I reached the fence on the far side of the reserve and stared down the dreadful plunge of Glaslyn's ravine, to the green pastures and woodlands far below. Clinging to the steepest slopes were a few young rowan trees. Otherwise the sides of the gorge were torn by erosion gullies. The bare rock and soil looked like the hills of Afghanistan. No crows or choughs winged the midway air; a solitary gull battled down the updraft towards the mild hedged fields of the South Dulas valley. The bitter, battering wind funnelled up the ravine and over the heath.

I strode back to the path which led to the lake. I soon found myself among a flock of sheep, grazing the low heather even lower. They stared at me as I passed, chewing, their white faces bland but oddly engaging. I resisted the urge, which always arises when I am watched by these creatures, to address them. I knew the question I wanted to ask: what are you doing here?

The lake was surrounded by a fine grey gravel that chinked like broken glass as I walked on it. Where the stones had been pressed into the peat by visitors' feet, powdery sage-coloured lichens crept over them. Wavelets rustled against the shore. There was not a tree or a shrub to be seen, except for the heather, which was nowhere higher than my knee. The reserve looked as brown and blurred as an old sepia photograph. It was a dismal place, almost as grim and almost as empty as the pastures around Llyn Craig-y-pistyll that I had visited the previous autumn.

There was a single clump of fern amid the heather. I saw one small heath butterfly – ginger and grey, furry, with a little black eyespot on the tip of its wing – pausing briefly on a tormentil flower. It was the only insect I would see on the reserve that day. The bilberry plants had been grazed almost to the roots. They carried no flowers or fruit: everything edible had been bitten off. Sheep's wool was dragged through the heather. But for two distant skylarks, an occasional pipit swooping away over the heath and the inevitable Canada geese on the lake, there was neither sound nor sight of any bird. The plants and animals of this jewel in the crown of the Cambrian Mountains were almost identical to the miserable remnants – the monotonous, impoverished moonscape left behind after the Atlantic rainforests had been destroyed – clinging to the rest of the wet Desert.

A small party of white-faced ewes lay on the gravel beside the lake in the sunshine, guarding the kissing-gate halfway along the shore. As I approached they hauled themselves to their feet and shoved their way through the heather to join a larger flock a few yards off. Some of them started rubbing themselves against the fence, rubbing off their scrappy, unshorn fleeces. Little tufts of wool clung to the knots in the wire.

The notice on the gate told me that 'Welsh white cattle are grazing this reserve'. I could not see them, but the land was overgrazed and poached: trampled, pitted and compacted. Here there was no heather, just grass eaten almost to the rootstocks, a few pillars of creeping thistle, their purple tips beginning to flower, and short thickets of soft-centred rush. It looked the same as any overgrazed pasture, yet this too was part of 'the wildest and most regionally important site' in Montgomeryshire.

On the leeward side of the lake the water stretched smooth before it shattered, a few yards from the shore, like a broken windscreen, into tiny fissures which extended into ripples then small waves on the far bank. The lake was perfectly clear. Its bed was covered with oddly regular chips of brown stone.

I crossed the cattle pasture and pushed on through denser thickets of rush, surrounded by brilliant green moss. I stumbled up the slope, which was soft and plumed with cotton grass. Everything in the nature reserve, it seemed, was below knee height, except the sheep and cattle. I jumped the fence, regained the track and followed it south towards Pumlumon. As it wound round a tump, I noticed, emerging from the heath on the far side of the site, two trees. I fixed my binoculars on them then swore out loud: Sitka spruce! The seed must have blown in from the great plantations across the mountains. They were, as far as I could tell, the only trees on the reserve, except for those clinging to the inaccessible slopes of the ravine. The white plague – or so I thought at the time – had destroyed the rest.

The track took me deeper into the Pumlumon site of special scientific interest, in which the Glaslyn reserve is embedded. When it crested a hill I found myself looking down on Llyn Bugeilyn, a sausage-shaped lake that filled a glacial valley. A distant raven planed into the wind. Above the lake was a ruined farmhouse. And there at last, growing within what would once have been the enclosures around the house, were trees.

I stepped into the broken-down barn at the back of the house, sat on one of the fallen stones and ate my lunch. Around me, bleached oak rafters were strewn over the ground. Ferns, willowherb and a small rowan rose from the stonework, out of reach of the sheep. Nettles had sprung up among the mossy, wormed beams. Around the base of the walls ground ivy and bittercress grew.

Overhanging the tumbled barn were a giant ash tree and an ancient mossy rowan, half toppled with age. Beyond them were smaller ashes, a stunted sycamore, then a short row of gnarled and ancient hawthorns, the last fragment of an old hedgerow. The sickly smell of their blossoms came to me on the wind in gusts. The ground beneath them was confettied with petals. Their roots writhed above the rocky turf, as if trying to force their way back into the soil. There was no sound but the wind in the trees.

I walked back into the moors. In brilliant sunshine I climbed Banc Bugeilyn, the hill overlooking the lake. It gave me an excellent vantage point. To the south massed Pumlumon: stubby, ragged in outline, pale khaki, like the rest of the land. I could see perhaps a quarter of the conservation area. I scanned the whole view carefully with my binoculars. I noted the clump of trees around the old farmhouse; one small cluster of sallow beside the lake; two more Sitka spruce trees and a few rowans clinging to a wall of the ravine too steep for the sheep to reach. Otherwise, the whole landscape, perhaps 2,000 hectares of this celebrated site, was treeless. Around me were signs of peat erosion caused by heavy grazing: little cliffs of black soil from which the surrounding bog had shrunk. Something had gone horribly wrong here.

I returned to the car, feeling empty and miserable. I turned on the ignition, removed the handbrake and set off down the road. After fifty yards I slammed on the brakes, parked as close to the edge of the narrow road as I could and jumped out again. I could scarcely believe what I had seen.

The sward on the verge was an exuberance of colours as rich as the Lord Mayor's Show. Here were drooping red spikes of sorrel, golden bird's foot trefoil like Quaker bonnets, the delicate umbels of pignut, heath milkwort – some pink, some blue – red campion and cut-leaved cranesbill. Here were little white flowers of eyebright, with egg yolk on their tongues, dark figworts, which released a foxy smell when I ran my hand through them, purple knapweed, pink and white yarrow, foxglove, mouse ear, male fern, deep cushions of bedstraw, wild raspberry, heath speedwell, hogweed and willowherb. Growing through the sward were little saplings of sallow and rowan.

A few hundred yards further along the road I stopped again. Taller rowans and sallows were growing on the verge, as well as hawthorn and elder. Around them the heather rose above my waist. The bilberries were covered in fat dark fruit and thick with cuckoospit. Small heath butterflies, little pale moths and chironomid midges swarmed around the plants. A bracken chafer in electric colours – a green iridescent head and thorax, bright copper elytra – crawled over the bilberry flowers, its strange three-fingered antennae sweeping this way and that.

This, I realized, was what I had seen in other parts of the mountains.

The only rich repositories of life were the verges of the roads, partly at least because the sheep could not reach them. (One experienced ecologist tells me that there could also be an effect caused by dust from the road fertilizing the verges. Another says that this would be more likely to reduce diversity than to increase it.) The highways authority, by ensuring that sheep are kept out of the traffic, has done more for nature conservation and biodiversity than the bodies charged with preserving our natural heritage. I thought that what I had seen in the Glaslyn reserve was a disgraceful failure, a shocking lapse of effective management. I soon discovered that it was worse than that.

Before I go further, I should say that I do not mean to single out the Montgomeryshire Wildlife Trust, which is run by devoted and conscientious people. As I will show in a moment, they claim that they do not have much choice over how they maintain their land. I have chosen this example not because it is exceptional but because it is typical: the trust's treatment of Glaslyn exemplifies the management of many nature reserves in the uplands of the United Kingdom.* Our national parks are in an even worse state. Foreigners often express their astonishment when they discover that many of them (ten out of the fifteen) are little more than sheep ranches, whose custody is almost indistinguishable from that of unprotected places. While Britain's is an extreme case, there are some aspects of this destructive form of conservation at work in other parts of Europe.

Soon after visiting Glaslyn, I read the trust's management plan for the reserve. To my amazement, I found that its grazed-out shell of an ecosystem, which can scarcely be distinguished from the rest of the Desert, has been deliberately kept like this. The plan seeks to ensure that the reserve remains in its current state: covered in close-cropped heather.† 'Invasive' and 'undesirable' species, it announced, will be removed.

---

* Among those which have, in my view, been kept by conservation groups in a similar state of desolation are Kielderhead, Whitelee Moor, Butterburn Flow, Harbottle Crags, Moor House-Upper Teesdale, Dove Stone and Geltsdale in England, the Isle of Rum, Rahoy Hills, Ben Mor Coigach, Cottascarth-Rendall Moss, Birsay Moors and the Oa in Scotland, Rhinog, Cwm Idwal, Cadair Idris, Y Berwyn and Yr Wyddfa in Wales, and Aghatirourke and Boorin in Northern Ireland.
† There should be no 'successional processes from upland heath to any other community'.[2]

What does this mean? I checked with the trust: invasive and undesirable species are native trees, such as rowan, sallow, birch and hawthorn, returning to their natural habitat. Even in the ravine, the plan insists, no more trees than already exist should be allowed to grow.

Another document published by the wildlife trust stated that cattle were to be kept on the grassy part of the reserve 'until there is an average sward height of 10cm'.[3] The trust revealed that 'to maximize the impact of the cattle, the grassland was strip-grazed'. This apparently, is how nature should best be protected in what this organization calls its 'flagship' reserve.[4] It is by these means that, at great expense, it sustains the ambience of a nuclear winter.

So why is this happening? The answer is like the Ouroboros, the snake swallowing its own tail. When you have followed it all the way round you find yourself back where you started.

The stated purpose of this brutal management regime is to maintain the heath and bare bog it contains 'in favourable conservation status' (it is failing dismally, but let us put that to one side for now). The plan points out that 'the site is artificial, having been created as the result of human activity following the removal of trees during the manufacture of lead'. It was kept treeless, before it became a nature reserve, by the farmers who burnt and grazed it. It must, the management plan insists, remain this way, fixed in time like the old sepia photograph it resembles. But nowhere is the obvious question asked or answered: why?

I had lunch in Welshpool with three people from the Montgomeryshire Wildlife Trust. I found to my surprise that they were in sympathy with much of what I said. So why were they managing the reserve like this? It was simple, they told me: that was the law.*

'We are given these targets and sites are designated for them. We're seriously in trouble if we don't abide by them. We wanted woodland to succeed naturally up the gulley [the great ravine at Glaslyn]. We want to fence it off and let it happen. But God have we had trouble.'

When I spoke to the chairman of the Countryside Council for

---

* 'If we don't abide by the criteria, we are breaking the law. We are told what the condition of the site needs to be. We're delivering exactly what we're obliged to do. There's no negotiation.'

Wales, which enforces the rules for managing the site, he disputed some of what the trust says,* but he agreed that some of the rules should be re-examined.

The owner of the land must keep its 'interest features' – particular plants and animals, habitats or geology – in 'favourable condition'.[5] The guidelines defining this are quite strict. In places like Glaslyn, for example, whose interest features include blanket bog and upland heath, they insist that scattered trees or scrub should cover less than a tenth of the bog and less than a fifth of the heath.[6]

These standards reflect European rules, which list the kind of places that countries must protect.[7]† Among them are wet heaths, moorgrass, blanket bogs and other such sheepwrecks, of the kind represented at Glaslyn.[8] One of the official reasons for choosing such places is that they are internationally important, because they possess 'assemblages of key species'.[9]

We have an international duty to preserve blasted heaths, bare bogs, acid grasslands and other such sheep-scorched places because they support a particular community of plants or animals or fungi or lichens. But every habitat – whether a rainforest or a railway track – supports a particular assemblage of species, a combination found nowhere else. The assemblage is a product of the physical habitat. By managing the land to protect one combination of species, we prevent other combinations from developing there.

For example, the display board at the entrance to the Glaslyn reserve explains that the species being protected there are red grouse, wheatear, skylark and ring ouzel. The land is managed partly to maximize their populations. But why? All four are close to the bottom of the list of species considered to be 'of European conservation concern'.[10] It is true that most are declining in the United Kingdom (the ring ouzel in particular), but that applies to many birds, plenty of which are in far greater trouble than these. In fact their relatively high numbers in Brit-

* Morgan Parry told me: 'We feel that we've had a fair degree of flexibility . . . I think the staff involved would say that they actually have come quite a considerable distance in terms of meeting objectives other than sustaining a barren upland environment.'
† The rules concerning the proportion of plants that may or may not grow in these places are national interpretations of the European rules.

ain and Europe are an artefact of grazing; they are all species which can survive in the scoured, open habitats humans have created and that some conservationists now seek to preserve, in order – with dizzying circularity – to protect the species which can survive here.

Of the four, the red grouse is the animal whose conservation best encapsulates the madness of current policy. It has no European conservation listing, because Europe contains so many of them. They are sufficiently abundant in Britain for thousands to be shot here every year and served, almost raw, in gentlemen's clubs and smart restaurants.

The number of red grouse in this country is sustained through the ruthless persecution of far rarer animals: the predatory birds and mammals that might reduce their numbers. So valuable is grouse shooting that even when this persecution is illegal and invokes (in theory, though seldom in practice) stiff penalties, it persists. Tests conducted by the Scottish government found that golden eagles, red kites, peregrines and a white-tailed sea eagle whose corpses were found on grouse-shooting estates had been poisoned.[11] Enough golden eagles were being killed to prevent Scotland's population (the only breeding population in Britain) from recovering. The white-tailed eagle was one of those reintroduced to Scotland at great expense and trouble, which have begun to establish a fragile clawhold on parts of the coast. One gamekeeper, on the Skibo estate in Sutherland, was caught in possession of enough carbofuran – a banned pesticide – to kill all the birds of prey in Scotland six times over. Three dead golden eagles were found on the estate; so was a dead grouse, pinned to a metal stake and saturated with carbofuram, which had evidently been laid out as bait. He was fined just £3,300.[12]

Red grouse are also maintained by a programme of cutting and burning which keeps the heather moorlands free from most other plants and ensures that there are plenty of young shoots for the birds to eat. This programme shuts out many of the other bird species which might have lived on the uplands.

So it is puzzling and disturbing to discover that the wildlife trust which manages the Glaslyn reserve describes red grouse as 'one of our key indicator species'.[13] An indicator of what? Its answer is 'the health of an upland landscape'. But what, in this context, does health mean? The red grouse is to the uplands what the magpie is to the lowlands:

it benefits from changes caused by humans. What is healthy for red grouse tends to be unhealthy for other species, even for other species of grouse, such as the black grouse, the capercaillie and the hazel grouse (which might have lived in Britain before we lost most of our forests). Sustaining the kind of habitat required to support artificially high numbers of red grouse destroys the habitat required by rarer species. So why are red grouse a 'key indicator'? Because they show the trust that the 'interest feature' – the treeless, blasted upland heath – has been maintained. We return to the head of the snake.

It is true that, unlike the red grouse, some of the species chosen as members of the favoured assemblages are rare. But some of those not chosen are even rarer: they no longer exist in many regions, because the habitats in which they lived have been replaced by the 'interest features' conservationists are trying to preserve. Both the wildlife groups and the official bodies are advised by ecologists. They defend the animals and plants they study as much for professional reasons as for environmental ones. Moorland weevil specialists become moorland weevil champions. A weevil ecologist tends to have little interest in capercaillie, and would respond with hostility to an attempt to expand capercaillie – or wildcat or lynx – habitat at the expense of weevil habitat. But because there are no longer any capercaillie, wildcat or lynx in Wales, and therefore no one studying them there, there is no competing group of local scientists arguing for capercaillie forests instead of weevil moor. Conservation policy is self-reinforcing.

There are two other official reasons for protecting particular places: 'high risk' and 'rapid decline'. These were the justifications the Montgomeryshire Wildlife Trust gave me for the way it manages its reserve. Heather, it said, 'is now a rare habitat with its distribution limited to Europe'.[14] This is questionable: there are between 2 and 3 million hectares of upland heath in the UK alone.[15] But why should the decline of a man-made habitat make it worthy of preservation? The contaminated land associated with active industry, fresh slag heaps and the tailings from deep coal mines are all in precipitous decline in Europe. If the criteria were to be applied even-handedly, these – and the sparse life they harbour – would be our conservation priorities.

Would it not be better to stop suppressing natural processes and allow the land to find its own way? Somewhere like Glaslyn is likely to revert to a mixture of rainforest, bog forest, scrub and heather. This would surely be a richer and more interesting place than the nineteenth-century ecological disaster being preserved there at the moment.

Some conservation groups claim that open habitats, with only scattered trees, represent the 'natural' state of the hills. They often call upon the work of Frans Vera, the man who founded the rewilding project in the Netherlands using Heck cattle. He has argued that the natural condition of most of the land in the warm, wet climate that has prevailed for the past 5,000 years is pasture with groves of trees. Grazing pressure by wild animals, he maintains, kept the forest open, much as sheep and cattle do today.[16] It is an interesting idea, but, overwhelmingly, the evidence does not support it.*

Others claim that the sheep or cattle or horses they keep on the land help to maximize the diversity of life. What they tend to mean is the diversity of certain kinds of life, such as butterflies or wild flowers: species which favour open, sunny places. But when you count species of all kinds – beetles, spiders, fungi, birds and everything else – native woodland turns out to be much more biodiverse than even the richest flowering meadows.† Most animals need places in which they can hide from predators, or which do not dry out quickly, or are protected from wind and sudden changes in temperature. Open landscapes tend to offer none of these defences.

---

* Tree pollen dominated the fossil record until people and their livestock began clearing the forests, suggesting that the land was mostly covered by deep forest.[17] Tree trunks found buried in bogs tend to be straight and unbranched, which suggests that they were competing for light with their neighbours.[18] Parkland trees, by contrast, branch close to the ground. The beetles which were abundant before the human population rose are the species associated with dense forest.[19] Even in previous interglacial periods, before massive or disruptive herbivores such as the mammoth, the straight-tusked elephant, the Merck's and narrow-nosed rhinoceroses, the hippopotamus and the water buffalo became extinct in northern Europe, the most widespread vegetation was closed-canopy forest.[20] Wild herbivores appear not to have been capable of creating the open landscapes Vera proposes.[21] While he argues that oak and hazel cannot grow in deep forest, there is plenty of evidence suggesting that they can and did.[22]

† Clive Hambler and Susan Canney note that 'Plagioclimax grasslands are often described as "species-rich", when in fact they are rich in flowering plants and are otherwise species-poor.'[23]

A study in the Cairngorms, in the Scottish Highlands, found that wooded habitats are eleven times richer in nationally important species than grassland, and thirteen times richer than moorland.* The figures are even starker when you consider creatures found nowhere else in Britain. There are 223 such species on the massif. One hundred of them are associated with woodland or trees. But just one – a fungus that lives on bilberry leaves – requires moorland for its survival. The management of upland nature reserves is informed by a profound misperception: that wildlife is best protected by clearing away the trees and scrub.

In one of its pamphlets, the Montgomeryshire Wildlife Trust warns that 'in some areas, heather moorland is declining in quality due to neglect of traditional moorland management techniques such as cutting and burning'.[25] Imagine how a tropical ecologist would respond if she saw that. British environmentalists have been campaigning for years to stop the cutting and burning of habitats in developing countries, yet here we see this destruction as an essential conservation tool. A conservation movement which believes that the environment is threatened by a lack of cutting and burning is one that has badly lost its way.

The choice of favoured ecosystems in this country and in some other parts of Europe appears arbitrary, guided by impulses which have been neither widely examined nor properly explained. The decisions we have made are historical, cultural and aesthetic, dressed up in the language of science.†

I would not object to this – the way in which we engage with nature will always be mediated by culture – were it not for the fact that some of the upland habitats we have chosen to conserve seem to me to be almost as dismal, impoverished and lacking in structure or complexity as a parking lot. This is not an entirely subjective view. Without trees, large predators, wild herbivores, rotting wood or many other components of a thriving ecosystem, these places retain only a few

---

* Despite being the main habitat for some 39% of important species, woodlands cover only about 17% of the land area of the Cairngorms. In contrast, moorland appears to support only 3% of the Cairngorms' important species, but covers some 42% of its area.[24]

† Some of them arise from Derek Ratcliffe's famous Nature Conservation Review in 1977, which identified semi-natural sites that he considered important for conservation.

worn strands of the complex web of life. The lively ecological processes I find so fascinating, the trophic cascades and unexpected
interactions, the constant surprises that in an untrained ecosystem
delight and enthral, are all prohibited.

These issues become still more pressing when you discover that, even
on its own terms, across much of the uplands this approach is failing
dismally. A survey of the birds in the Pumlumon site of special scientific
interest, of which the Glaslyn reserve is part, found that there had been
a catastrophic decline in the species the severe regime is supposed to
protect.* Their numbers, the survey found, have been falling at greater
rates inside the conservation area than in Wales as a whole. Extreme
management is not working, even by the standards it sets for itself.

Ecologists profess themselves mystified by this failure. It could be that
the management programme simply cannot sustain the species it is
designed to protect, as the sheep it relies on gradually degrade the habitat: the longer they stay there, the more damage they do. Their compaction
and poaching of the land, for example, could reduce the number of lar-
val insects on which many birds depend. Or it could be that trying to
preserve the ecosystem as if it were static prevents it from adapting to
changing conditions such as global warming and acid rain (which is still
an issue in these very wet places). All we can say at this stage is that the
current conservation model appears to have failed. In its management
plan the wildlife trust remarks that the habitats it has been trying to
preserve since 1982 at Glaslyn remain in 'unfavourable condition'.[28] The
same certificate of failure has now been issued to 60 per cent of the most
important wildlife sites (the special areas of conservation) in Wales.[29]

Some people have responded to such failures by blaming the fact that
the habitats they have been saving are too small. The answer, they say,

---

* Between 1984 and 2011. At the top of Glaslyn's plan is a list of the eight birds for
which the reserve is considered 'very important'.[26] One of these, the short-eared owl,
did not appear in either the 1984 or 2011 surveys. One other, the hen harrier, rose by
a single nesting pair: none were seen in 1984; one was spotted in 2011. One pair of
peregrines was seen in both cases. The rest were in freefall. Red grouse, skylark and
wheatear had all declined by around 50 per cent. The golden plovers seen in 2011 had
fallen by 92 per cent. Ring ouzels were not found by the second survey at all. The
report notes that 'large scale declines across nearly all the species that occur on the site
were recorded'.[27]

is to move towards 'landscape-scale' conservation: doing the same thing across a wider area.* But surely the problem is not only size but also method? That intensive management, sooner or later, will fail? If for no other reason, this will happen as temperatures rise. Locking in particular assemblages of animals and plants will become ever less viable as conditions change. If an ecosystem cannot adapt, its richness, structure and complexity will decline even faster than they are declining today.

The plan for Glaslyn claims that 'wider knowledge of the Trust's work and the rationale behind management will create a more sympathetic public'. I suggest that if people better understood its work and rationale, it would have the opposite effect.

The promise of conservation used to be that by protecting the species you would protect the habitat. The Bengal tiger needs jungles to survive, so defending it means defending the rich and fascinating ecosystem that supports it. But in the United Kingdom, the species we have chosen, historically, to protect are often those associated with damaged and impoverished places, and to defend them we must keep the ecosystem in this state. Armies of conservation volunteers are employed to prevent natural processes from occurring. Land is intensively grazed to ensure that the plants do not recover from intensive grazing. Woods are coppiced (the trees are felled at ground level, encouraging them to resprout from that point) to sustain the past impacts of coppicing. In their seminal paper challenging the conservation movement, the biologists Clive Hambler and Martin Speight point out that while coppicing might favour butterfly species which can live in many habitats, it harms woodland beetles and moths that can live nowhere else.[30] They noted that of the 150 woodland insects that are listed as threatened in Britain, just three (2 per cent) are threatened by a reduction in coppicing, while 65 per cent are threatened by the removal of old and dead wood. (This is not to suggest that coppicing has no ecological role: many woodland species must have evolved to take advantage of the habitat disturbance caused by elephants.)

Conservationists sometimes resemble gamekeepers: they regard some of our native species as good and worthy of preservation, others as bad and in need of control. Unlike gamekeepers, they don't use the

---

* This roughly speaking, is the approach of the celebrated report by Sir John Lawton.

word 'vermin' to describe our native wildlife. Instead they say 'unwanted, invasive species'. They seek to suppress nature, to prevent successional processes from occurring, to keep ecosystems in a state of arrested development. Nothing is allowed to change: nature must do as it is told, to the nearest percentage point. They have retained an Old Testament view of the natural world: it must be disciplined and trained, for fear that its wild instincts might otherwise surface.

The result is back-to-front conservation. Wildlife groups seek to protect the animals and plants that live in the farmed habitats of the previous century, rather than imagine what could live there if they stepped back. They take a species like the red grouse, or a club moss or a micromoth, which happens to thrive in a place that has been greatly altered by humans, and they build their management plans around it, seeking to keep the land in the state which best secures its survival. In doing so, they shut down the opportunities for other species to establish themselves, either naturally or by reintroduction.

Sustaining the open, degraded habitats of the uplands means keeping sheep. It does not seem to matter whom you talk to in the hilly parts of Britain: farmers, government officials and wildlife groups will all tell you that the answer is sheep – what was the question? If you challenge their management of the land they invariably invoke the horror of 'undergrazing'. But how can a native ecosystem be undergrazed by a ruminant from Mesopotamia? Is our wildlife under-hunted by American mink? Are our streamsides under-colonized by Himalayan balsam, our rivers under-infested by red signal crayfish, our verges under-occupied by Japanese knotweed? It is a nonsensical concept.

Even the grazing of cattle or horses in the uplands, which some conservation groups characterize as the benign alternative to sheep, means maintaining habitats that would not exist without us. During the Boreal and Atlantic periods, when warm, wet weather returned to northern Europe, the giant aurochs, or wild cow, appears to have been a forest animal. Analysis of the carbon and nitrogen isotopes in its bones shows that it lived on woodland plants. Domestic cattle, by contrast, from their first appearance in northern Europe, largely ate grass, growing in clearings created by people. The chemical differences are so discrete that they can be used to distinguish the bones of wild cattle from the bones of domestic cattle.[31]

The wild horse seems to have disappeared from the British Isles around 9,000 years ago – some 2,000 years after the last ice sheets retreated.[32]* Though hunting by humans doubtless accelerated its extinction, the horse was deprived of what was likely to have been its favoured habitat – steppe grasslands – by the change in climate, which allowed forests to spread. In other words, the horse died out here soon after the lion[33] and the saiga antelope[34] and before the reindeer.[35] Though both horses and aurochs were intensively hunted, the aurochs survived for much longer: until 3,500 years ago in Britain and into the seventeenth century on the Continent. This is one of several lines of evidence suggesting that climate change, not hunting, was the major reason for the horse's disappearance.[36] Arguably, it no more belongs to our native fauna under the current climate than the woolly mammoth does. The large herbivore which *is* missing from our ecosystem is the moose or elk (*Alces alces*), which became extinct here a little under 4,000 years ago, largely as a result of hunting.[37] Moose are browsing animals which live in and around forests.

But even if horses or cattle were replacing native plant eaters, the absence of predators utterly changes the way in which they engage with the ecosystem. The grazing regime imposed by conservationists in upland Britain – whether they are using sheep, cattle, horses, yaks or pushme-pullyous – bears no relationship to anything found in nature.

What we call nature conservation in some parts of the world is in fact an effort to preserve the farming systems of former centuries. The idealized landscape for many wildlife groups is the one that prevailed a

---

* There are two references to horse remains beyond this date in the archaeological record. One, found in Kent and held by the Harrison Institute, is sometimes described as being 8,000 years old. I checked with the institute: it appears that some people had confused BC (Before Christ) with BP (Before Present). This institute tells me it has been carbon-dated at around 9,760 years old. The other, a single tooth, was found in a Neolithic tomb at Hazleton in Gloucestershire, which is some 5,700 years old. In correspondence with myself and the biologist Clive Hambler, Robert Hedges, one of the archaeologists who analysed the contents of the burial site, explains that the tooth itself is undated and the notion that it originated at the same time as the tomb is 'an unsupported possibility only'. It is possible that it was found and carried into the tomb by Neolithic people. If horses had survived that long in Britain, one would expect to see a good deal more fossil evidence, before they returned in domesticated form, later in the Neolithic.

hundred years ago, regardless of the point at which they start counting. This is what they try to preserve or re-create, defending the land from the intrusions of nature. Reserves are treated like botanic gardens: their habitats are herbaceous borders of favoured species, weeded and tended to prevent the wilds from encroaching. As Ritchie Tassell says sardonically, 'You wonder how nature coped before we came along.'

I do not object to the idea of conserving a few pieces of land as museums of former farming practices, or of protecting meadows of peculiar loveliness in their current state, though I would prefer to see these places labelled culture reserves. I do not object to the continued existence of reserves in which endangered species which could not otherwise survive are maintained through intensive management.* Nor do I believe that rewilding should replace attempts to change the way farms are managed, to allow more wildlife to live among crops and livestock: I would like to see that happen too. But if the protection of nature is to be extended to wider areas, as both conservationists and rewilders agree that it should be,[39] I believe we should first conduct a radical reassessment of what we are trying to achieve and why.

This assessment is likely to show us that rewilding could offer the best chance of protecting endangered species. According to a paper in *Biological Conservation*, around 40 per cent of the creatures that have become extinct in Britain since 1800 lived in woodlands, and two-fifths of those needed mature trees and dead timber to survive. The paper warns that 'extinction rates in Britain will rise this century without . . . restoration of woodlands and wetlands'.[40]

A new assessment might prompt conservationists to focus less on species and habitats which happen to be there already, and more on those which could return. Rather than sustaining the sheepwrecked, open habitats of the uplands, they might begin to reduce the impacts of human management, to allow trees to return, even to reintroduce some of the great beasts which once lived among them. That, to me, is a more inspiring vision than sustaining a slightly modified version of the farming which is suppressing the natural world almost everywhere. Everyone should have some self-willed land on their doorstep.

---

* Hambler and Canney argue that rewilding protects a greater number of threatened species than any other approach.[38]

Attitudes are slowly beginning to change. The Countryside Council for Wales talks about allowing 'a more natural cycle of growth and succession', and letting plantlife 'develop to its full potential'.[41] When I interviewed its chairman, Morgan Parry, he told me: 'I would agree that another world is possible and more desirable ... I would like to think that we can open our minds to the possibility that other landscapes can exist and they don't necessarily need to exist because of farming.'

He acknowledged that the idea of keeping the uplands open and treeless 'does need to be challenged'. So do the rules: 'I'm very supportive of thinking about how we might move towards a less predetermined outcome.' But the change, he said, cannot come from governments and their agencies; it is up to campaigners to mobilize public opinion to make it happen.

In a few places, something resembling rewilding is beginning, slowly and uncertainly, to happen. At Ennerdale in the Lake District, the National Trust, the Forestry Commission and a water company are granting nature a kind of day release from the conservation prison. It's a good start. But – apparently because they do not want to offend or frighten local farmers[42] – they cannot quite bring themselves to keep agriculture out of it, and insist on running some cattle on the land.

In parts of Essex and Suffolk, fields are being allowed to revert to saltmarsh, partly to protect the coast from erosion and storm surges. The transformation happens at great speed: after just a few years of inundation, rewilded barley fields support samphire, mullet and flounder, crabs, clams and flocks of wading birds.

In the lowlands of eastern England, government bodies and a wildlife trust have started what they call the Great Fen project, allowing some of the old peat fens to flood. It is not quite rewilding and is still informed by the curatorial ethos. It differs from usual conservation practice in that it is trying to re-create the landscape not of 100 years ago but of 400 years ago: a mixture of grazing land, reedbeds, woods and bogs.[43] The people running it hope that birds such as spoonbills and cranes will return. There are several dozen similar projects in Britain, many of them hybrids between conservation and rewilding, allowing nature more freedom than before, but in most cases unable to kick the addiction to livestock and management.

Even by European standards – let alone those of North America and much of the rest of the world – the United Kingdom has a peculiar fear of nature, and its conservationists a peculiar fear of letting go. Germany, France and Slovakia are permitting part or all of their national parks to rewild. Most countries in Europe now have large areas of self-willed land.[44] Even the tidy, busy Netherlands is allowing nature to reassert itself. But we remain, as a Francophone woman I know once rudely remarked about British men, '*constipé et embarrassé*'.

It need not be like this. I am convinced that before long it will cease to be like this. Conservationists will begin to ease their grip on the natural world. Some of them, I have discovered, are almost ready to do so. A change is on the way, which could start to transform places that now seem bleak and almost dead into a rich and complex ferment of life.

13

Rewilding the Sea

The Kraken sleepeth: faintest sunlights flee
About his shadowy sides; above him swell
Huge sponges of millennial growth and height;
And far away into the sickly light,
From many a wondrous grot and secret cell
Unnumbered and enormous polypi
Winnow with giant arms the slumbering green.

Alfred Lord Tennyson
The Kraken

They had been spotted two days before, on the edge of the reef at Llansglodion. The migration had begun; these were the scouts. Soon the rest would arrive: in battalions, divisions, armies, so many that you could scarcely put your foot down for fear of treading on one. Then, in a fortnight, they would be gone. Later in the year their ghostly husks would litter the beaches. A day as calm and warm as this could not be wasted.

The oaks had put out embryo leaves as minutely serrated as mouse paws. The fronds of the horse chestnuts in town, which had hung like empty gloves, began to stiffen and splay. Bracken unrolled leaflet by leaflet like a Mandelbrot set. On Llansglodion beach I glanced at the dismal seafront – the peeling guesthouses in hangover colours, faces shut to the sea, shops and houses in one hundred shades of grey and beige, their drabness accentuated by gaudy ice cream signs – then turned to face the other way. It was half an hour before low tide. The sea had retreated far beyond the breakwaters, and the bottom half of the beach shone like a mirror in the hazy sunshine. The bay opened into a long shallow crescent. In the north the dim hulks of Pen Lleyn and Ynys Enlli hunched above the horizon; in the south Pencaer – Strumble Head – sat like a low cloud on the water. The sea gleamed rhenium, embossed with dark bands as the waves rolled in.

I pulled on my winter wetsuit and a hood, and clambered over the rocks on the edge of the beach, slithering on wrack and gutweed. On the far side of the reef I met a man I knew, up to his waist in a rock-pool, netting prawns for bait. Yes, he said, they were here. I clamped my mask and snorkel to my face and slipped into the sea. The water's cold fingers crawled under my suit and down my back.

Where the waves had churned up the mud beside the rocks, the sea was opaque, so I struck out into the clearer water beyond. I could hear my breathing resound in my head, loud and hollow. I could just see the bottom and the dim pale flecks of shells on the mud. I pushed out further, enjoying the power that comes from swimming with your head down: it felt as if my arms had grown. When I raised my face, I found that I had started swimming back towards the rocks.

I set off again, put my head down and saw something that looked like the kind of exotic weapon that might be discovered during a raid on the home of a martial arts fanatic. The water was too cloudy to tell how deep it lay or how large it was. In the olive gloom it could have been a mile down, a benthic monster prowling the fringes of the contin-ental shelf. It was bunched up as if ready to spring, a snarl of spikes and legs and latent power. I was not wholly sure that I wanted to meet it.

I filled my chest with air and duck-dived to the bottom. I had no flippers, and the seabed, perhaps two and a half fathoms down, was at the limit of my dive. I touched the beast. It raised its long pincers over its back. I ran out of air and corked to the surface. I tried again, too quickly, knowing that I could lose the mark in the soupy water. This time I managed to get one hand beneath it. But its feet were planted in the seafloor and I had to surface before I could lever it up. Over-eager, forgetting myself, I took another great lungful and plunged back down. I grabbed it with both hands then kicked up, using my buoyancy to lift the creature. I was astonished by its weight. I reached the surface and drew a breath so sharp that it pulled down the stop valve. I tried again with the same result, and nearly asphixi-ated. I spat out the snorkel and took in a mouthful of water. Almost panicking, I put my head back, wheezing, coughing up brine. Yet still I would not let go of what I had caught. I clutched it to my chest with one hand even as I struggled to stay afloat. Evolutionary biologists have identified a rule they call the life/dinner principle. A predator puts less effort into the chase than its prey: if the hunter fails it loses only its dinner, if the hunted fails it loses its life. In this case the equation was reversed.

Breathing raggedly, lying on my back, I kicked towards the rocks until the water was just shallow enough, with my chin raised, for me to stand. I tiptoed to the reef, slithered over the weed and sat on a boulder, still panting, still pressing the creature against my wetsuit. I lowered it onto the rock and studied it. It looked like the grab used to lift crushed cars in a scrap-metal yard. Its claws were more than two feet from tip to tip, powerfully ridged and bossed, crenallated on the cutting edges. Every leg ended in a long black spike, which it had used to embed itself in the mud when I had tried to lever it out. Now the monstrous spider crab curled up and played dead. The only movements I could see were the bubbles which fizzed and popped from under its carapace.

Its shell was covered in weed and sponges: it had not yet moulted. It bulged with the suggestion of muscle like a Roman suit of armour. It was guarded with gothic spines and pinnacles, each surrounded by a ring of short bristles, and fringed with spikes like the Statue of Liberty, extending between the eyes into a pair of horns. The underside of the monster was covered in smooth articulated plates. It looked like a rock that had crept into life. Beneath its robot joints, its mineral crust, it scarcely seemed animate. I thought of these heavy creatures trundling out of the depths at the end of winter, slowly converging on the shore, and wondered what, among the disaggregated ganglia that pass for the crustacean brain, they perceived, what spirit moved beneath the expressionless shell. It was a male, which meant that I could keep it.

I travelled up the coast to look for clear water. Two miles to the north of Llansglodion the dunes billowed onto clean sand. I walked down the beach and into the sea. The water was bright enough to let the sun sparkle on the seabed. There was no chance of losing my way here: the ripples on the seafloor ran north to south, and as the waves rolled in towards the shore they knocked little puffs of sand east-wards. Head down, I could give myself up to that world.

It belonged to the crabs. Hermit crabs, helmeted in cowls and spires – winkles, turitellas, dogwhelks and topshells – scuttled over the seabed close to the beach, top heavy, almost upended by the pass-ing waves. As I moved into deeper water, they ceded the ground to masked crabs, the size and shape of bantam eggs, whose pincers, like articulated forceps, were twice the length of their bodies. I watched one stuffing a smashed shellfish into its mouth. Shore crabs in pie-crust shells scuttled away as I loomed overhead.

The tide had been rising for an hour and a half. I swam towards the horizon, feeling the cool green water push past my face. Creeping over the sand in two fathoms of water was a pink grapefruit carapace. I dived and swept it up in one movement, almost piercing my hands on the spines. It was a female – I let her go again. She drifted back to the seabed, paddling a little to keep her balance. I swam on and soon, in deeper water, spotted a much larger beast. I hung above it, feeling like a hawk about to swoop on its prey. When I had gorged on air, I dived. I needed both hands to lever it out of the sand. It was another male, the same size as the monster I had caught in Llansglodion.

I left it in a beach pool and swam out again, porpoising through the water, thrilled by the cold draught of the sea and the beams of light that searched the green deeps, glittering with motes of sand, drawn from the shore until I could no longer see the seabed, then into the emerald water beyond. I swam until my hands became so cold that I could not close my fingers. Even then I was reluctant to leave. My skin when I stepped out of the sea was white and riven.

I fished crabs on three more occasions that fortnight, while the wea-ther held, and watched as their numbers rose until they piled against the shore like autumn leaves. As they converged on the beach, I was soon able to pick them up from the undersides of the rocks at the bot-tom of the tide, without venturing into the water. Just beyond this mark, in the dull yellow light behind the breaking waves, they loomed through the wrack like armoured spaceships. Their flesh was sweet and firm, cleaner than crayfish, more tender than lobster. A large crab would feed three people. At the end of May, they disappeared as sud-denly as they had arrived. Later in the summer their cast-off shells washed up in crisp pink drifts sometimes a mile or two long, a last gift to the earthlings as they lumbered back into the deeps.

Older people I know, who have lived on this coast since they were children, told me that the spider crabs started arriving in large numbers only fifteen or twenty years ago. 'It's an invasion,' said the man who runs the tackle shop in Llansglodion. Some people assumed that they were moving north as the sea warmed. This is possible, as the species is limited by temperature: the hard winter of 1962–3 wiped out the spider crabs from the south-east of England. Others suggested that the disappearance of fish, which eat and compete with crabs, has allowed their population to explode. Something like this happened on the Grand Banks off Newfoundland, where crabs and lobsters proliferated after the cod were fished out.

Whatever the explanation may be, this migration is a reminder of a natural abundance that was once universal. I have seen spider crabs described as 'the wildebeest of our waters',[1] but there is, or was, nothing remarkable about their numbers. Almost every ecosystem – whether on land or sea – once resembled the Serengeti: great herds of animals, coming and going in prodigious migrations. The state of nature is a state of almost inconceivable abundance.

In his magnificent but sadly neglected book *The Unnatural History of the Sea*, Professor Callum Roberts recalls the herring migrations that once stormed the coasts of Britain.[2] Some shoals, he estimates, 'could block the light from 20 or even 40 square kilometres of seabed'. He quotes Oliver Goldsmith who, in 1776, described the arrival of a typical body of herring 'divided into distinct columns, of five or six miles in length, and three or four broad; while the water before them curls up, as if forced out of its bed . . . the whole water seems alive; and is seen so black with them to a great distance, that the number seems inexhaustible'.[3]

Goldsmith noted how these shoals were harried by swarms of dolphins, sharks, fin and sperm whales, in British waters, within sight of the shore. The herring were followed by bluefin and longfin tuna, blue, porbeagle, thresher, mako and occasional great white sharks, as well as innumerable cod, spurdog, tope and smoothhound. On some parts of the seabed the eggs of the herring lay six feet deep.

Even within the past century such monsters as pursued those shoals were still circumnavigating our coasts. Bluefin tuna, sometimes described by fishermen hunting pilchard and herring as 'blue mackerel' or 'king mackerel', roamed through all the seas around Britain. As the angling expert Mike Thrussell records, in the late 1920s big-game hunters heard tales of vast fish appearing among shoals of herring in the North Sea. In 1930, fishing off Scarborough on the Yorkshire coast, they landed their first five fish, all between 400 and 700 pounds.[4] By 1932, in the same waters, they had beaten the world record for bluefin tuna. They did it again in 1933, with a monster of 850 pounds. Some remarkable footage of these early expeditions exists. Tweedy men and women, angling from a tiny launch, used split-cane rods and gearless reels to catch this king of fish. One shot shows nine monstrous tuna lying on the deck of the steam-trawler the anglers used as their mother ship.[5]

Perhaps this contributed to their decline. The industrial fishing of herring and mackerel after the Second World War must also have done so, and by the late 1950s the sport-fishing had ceased. Since then the odd fish has been taken in nets, including one, off the Irish coast, of over 1,200 pounds.

The migrations inland were no less impressive. Before they were silted up by forest clearance and the runoff from ploughing, before they were weired, impounded and polluted, the water in most of the rivers in Europe is likely to have been clear. Most rivers would also have supported runs of migratory fish on the scale of those the Europeans encountered when they first arrived in North America. There they found sturgeon, some of them eighteen feet long, moving up the rivers in such numbers that, an English visitor recorded, 'in one day, within the space of two miles only, some gentlemen in canoes caught above six hundred ... with hooks, which they let down to the bottom and drew up at a venture when they perceived them to rub against a fish'.[6] Which river was this? The Potomac, that foul drain which runs through what is now Washington DC.[7]

Above the bottom-hugging sturgeon, Callum Roberts tells us, swarmed alewife and shad (migratory members of the herring family) in such numbers that there seemed to be more fish than water: in 1832 European settlers caught almost 800 million of them in the Potomac alone. In other rivers salmon were packed so densely that, an English army captain remarked, a gun could not be fired into the water without hitting some of them.

Oysters formed reefs across the bays and rivermouths that presented a hazard to shipping. The colonists, one source claims, picked twenty-pound lobsters out of the rockpools – and could think of nothing to do with them except use them as bait or feed them to the pigs.[8] Fathom-long halibut were caught only for their heads and fins; the flesh was discarded as inferior to that of other species that thronged the coasts.[9]

There is no reason to believe that the volume of fish and shellfish would, in an undisturbed system, have been any lower in Europe. We are less aware of what went before only because humans reached Europe earlier, their later technologies were more intrusive and their harvests more intensive than those of the Native Americans. The decline of the great herds of the rivers and seas began, in many cases, long before it could be recorded in writing.

But early documents hint at what there once was. Shoals of migratory fish whose existence in Britain we have all but forgotten jammed the rivers: shad, lamprey and sturgeon, jostling with hordes of salmon and sea trout. Until the eleventh century, when the diet shifted to marine fish, probably as a result of the depletion of freshwater species, they helped to feed much of Britain. By the thirteenth century sturgeon were so rare that only the king was permitted to eat them. But the marine ecosystem, when large-scale exploitation began, must still have been close to the opulent state early travellers later encountered in the New World. Roberts reports that Viking settlements in the north of Scotland were characterized by a mass of remains of cod, pollock and ling much bigger than any caught in inshore waters there today.

Everywhere the animals that lived in the sea were both more numerous and bigger than they are today. Cod commonly reached five or six feet in length. Even the great white shark is not as great as it once was. Roberts tells us that 'today, the maximum length of a great white shark is listed in guidebooks as 6 metres, but reports in the eighteenth- and nineteenth-century literature, too numerous and detailed to be dismissed, suggest sizes of 8 or 9 metres were not uncommon. Accounts at the time compare them in size with whales.' Haddock were once a yard long. Plaice were the size of road atlases, turbot like tabletops. The specimens we see on the fishmonger's slab are, for the most part, youngsters, caught before they were able to reach even a tenth of their maximum weight.

Genetic profiling of the great whales suggests that their populations, before whaling began, were higher than biologists had assumed. The larger the original population, the greater the variation within that which remains. An analysis of genetic data in the journal *Science* suggests that the North Atlantic alone supported around 265,000 minke whales, 360,000 fin whales and 240,000 humpbacks.[10] Today the minke whales, after a severe population decline, have recovered to 149,000, the fins to 56,000 and the humpbacks to 10,000. The whales once visited all the seas of the region; by the eleventh century they were being hunted in both the English Channel and the North Sea.[11]

Just as on land, the ecology of the sea is more complex than scientists once assumed. The trophic cascades now being discovered in the oceans are, if anything, even more remarkable than those of the terrestrial ecosystem. Fishermen and many fisheries scientists, for example, have long assumed that if whales are removed from the southern oceans, the volume of their prey – mostly fish and krill – will rise. This argument has been used by the Japanese government to justify its continuing slaughter of these beasts.[12]

But recent work suggests that reducing the population of whales might have had the opposite effect. As whale numbers have declined, so have the krill:[13] to just one-fifth of their volume before the 1980s.[14] Their collapse, until recently, mystified observers. It now seems that the whales perform an essential role in keeping nutrients in the surface waters. If undisturbed, the plant plankton at the bottom of the foodchain sinks out of sight, beyond the photic zone (the waters in which the light is strong enough to permit plants to grow). The nutrients it contains sink with it, becoming unavailable to most lifeforms. The surface waters rapidly become depleted of essential minerals, especially iron, whose scarcity limits growth. In the summer, when plant plankton is reproducing fastest, the wind and waves drop, allowing it to sink more rapidly. The same applies to the faeces of the animals that eat it.

Even today, a study in the journal *Nature* calculates, the mixing power caused by movements of animals in the oceans is comparable to that of the wind, waves and tides.[15] This, it says, is a conservative estimate. When whales were more abundant, the effect would have been still greater. Simply by plunging up and down through the water column, the whales help to keep plankton circulating in the surface waters. But their impacts extend far beyond that. They often feed at depth and defecate at the surface, producing great plumes of iron-rich manure that fertilize the plants in the photic zone, on which krill, fish and other animal plankton feed. One paper estimates that, before their population was reduced, whales recycled at least 12 per cent of the total iron content of the southern ocean's surface waters.[16] More whales meant more nutrient cycling, which gave rise to more plankton, producing more fish and krill.

Another study, in the Gulf of Maine, estimates that whales and seals, by defecating at the surface and recycling nutrients there, would, before they were hunted, have been responsible for releasing three times as much nitrogen into those waters as the sea absorbed directly from the atmosphere.[17] Whales in the gulf typically dive to a hundred metres or more to feed, bringing back the nutrients they harvest to the surface. The volume of plant plankton has declined across most of the oceanic regions in which it has been studied over the past century. The principal reason is the rising temperature caused by man-made climate change.[18] But according to the marine biologist Steve Nicol, the decline has been steepest where whales and seals have been most heavily hunted.[19] The fishermen who have insisted that the predators of the species they hunt be killed might have been reducing, not enhancing, their catch.

If the production of plankton declines, so does the transport of carbon to the deep ocean. By stimulating plankton blooms through recycling iron, another study suggests, sperm whales in the southern oceans cause the removal of around 400,000 tonnes of carbon from the atmosphere every year.[20] The extra plants absorb carbon dioxide, then, after being kicked around the surface waters a few times, sink into the abyss, where the carbon remains for a very long time. The whales also release around 200,000 tonnes of carbon through respiration, which means that, on balance, roughly the same amount of carbon is taken out of circulation.

When you consider that the sperm whale is just one of several species, the southern ocean is just one of several regions and the current number of leviathans is a fraction of what it once was, it becomes clear that whales could once have caused the sequestering of great quantities of carbon, perhaps tens of millions of tonnes every year. This is enough to make a small but significant difference to the composition of the atmosphere. Another paper maintains that during the twentieth century the whaling industry shifted over 100 million tonnes of carbon from the oceans to the atmosphere, simply by turning whales into oil and other products that were burnt or otherwise oxidized.[21] Allowing whale numbers to recover could be seen as a benign form of geo-engineering.

The removal of the great sharks, which took place, on the whole, later than the destruction of the whale population, has had similarly devastating effects. Caught for their fins or accidentally by nets and lines set for other species, big sharks have vanished with astonishing speed. Off the eastern seaboard of the United States, for example, in the thirty-five years beginning in 1972, tiger sharks declined by 97 per cent, scalloped hammerheads by 98 per cent and bull sharks, dusky sharks and smooth hammerheads by 99 per cent.[22] The result is an explosion of animals which no other species is big enough to eat: large rays and skates and smaller sharks. Many of them have increased tenfold or more. In Chesapeake Bay alone, for example, there are now an estimated 40 million cownose rays.

Cownose rays eat shellfish, and this population consumes some 840,000 tonnes a year – almost 3,000 times as much as the total landing of clams of all descriptions in Virginia and Maryland.[23] By 2004 they had wiped out North Carolina's scallop fishing industry and were rapidly doing the same for oysters, hard clams and softshell clams. The economic damage caused by the destruction of large sharks surely outweighs any money made by catching them.

The collapse of the cod shoals off north-eastern America has had the opposite effect. Released from their predators, commercially valuable shellfish – in this case, shrimps, crabs and lobsters – have exploded, creating a new industry as valuable as the one it replaced. These too are now being heavily exploited.[24] Regardless of the economic consequences, the destruction of one of the world's great natural spectacles – the vast spawning aggregations on the Grand Banks and other shallow seas off the Atlantic coast, and the frenzy of tuna, sharks, dolphins and whales attendant upon them – is a tragedy.

In some places where cod were abundant they have failed to return even when fishing for them has ceased. This could be because cod appear to engineer their environment, creating the conditions necessary for their survival. On the Grand Banks they preyed heavily on mackerel and herring. When most of the cod disappeared, the population of mackerel and herring boomed, with the result that the relationship was reversed. The smaller fish became major predators of cod, eating their eggs and fry before they could mature.[25] The same thing has happened in the Baltic Sea, where cod eggs are eaten by herring and sprats.[26]

Turtles also appear to have changed the world to suit themselves. When Columbus arrived in the Caribbean, according to one study, that sea contained 33 million green turtles.[27] There were similar concentrations off the east coast of Australia and in other tropical and subtropical seas: turtles all the way down. Today there are 2 million green turtles, worldwide. They largely subsisted on turtle grass, a weed which once grew on the beds of great tracts of shallow water. These were the savannahs of the sea, supporting vast herds of grazing animals: dugongs, manatees and herbivorous fish as well as unimaginable numbers of green turtles (which were, permitted to live into old age, much larger than the average size of those of today). The grazers, in turn, supported marine lions, hyenas and cheetahs: big predatory fish, mammals and in some places reptiles, namely giant saltwater crocodiles.

When the turtles were slaughtered, mostly before the nineteenth century, the remaining population could no longer keep the turtle grass cropped. As the blades grew longer, they shaded the seabed and shielded the sediments from the current. The weed, uneaten, started to age and rot, and detritus built up in the still water beneath the beds. This became a food source for parasites which then began to destroy the living grass (a process biologists call 'turtle grass wasting disease'). Across much of the range that green turtles once occupied, the turtle grass has died off.[28] This, in other words, is a similar story to that of the mammoth steppes of Beringia, which, as the grass grew longer and its detritus insulated the soil, turned to mossy tundra when the grazing animals were killed (see chap. 6).

Perhaps the most famous trophic cascade in the seas took place along the eastern rim of the Pacific, where sea otters, once widespread and abundant, were almost wiped out by both native people and fur traders. The result was the near-disappearance of the coastal ecosystem. Sea otters prey on urchins among other species. Sea urchins graze on kelp, the long and leathery seaweed that, in the right conditions, produces tall, dense growths reminiscent of terrestrial forests. These harbour a wonderful variety of fish and other creatures. When the sea otters were killed, the urchins wiped out the kelp forests, bringing down the rest of the natural system.[29] In the few places in which the otters have survived and begun once more to proliferate, the kelp forests have started to return, just as the reintroduction of wolves to the Yellowstone National Park has permitted the trees to grow back. But now, in one of their remaining strongholds, the Aleutian archipelago, the sea otters are disappearing again, apparently because of another disruption of the ecosystem. Killer whales, deprived by human hunters of the seals and sealions they once preyed upon, have started eating the otters instead.[30]

Fishing has transformed the life of seas everywhere, to a much greater extent than most people know. As on land, no removal of an abundant species is without consequences, consequences that often ramify through the system. Take the humble oyster. I have mentioned the remarkable abundance of oysters on the eastern seaboard of the Americas that the first European adventurers encountered. It appears that similar concentrations were once found in other seas. A map made in 1883, 500 years after trawling began there, marks an area of the North Sea the size (inevitably) of Wales as oyster reef.[31] Before the age of trawling and dredging, it is possible that most of the North Sea bed was encrusted with oysters, while shellfish of other species would have colonized the sediments on which oysters could not settle.

One result is that this grey sea might once have been clear. Like most two-shelled molluscs, oysters filter the seawater. They also stabilize the sediments of the seabed. Less mud would have been raised, and that which was washed into the water would quickly have been extracted again. As the great beds were smashed by fish trawlers and oyster dredgers, the sea's filters were shut down at the same time as the crust of life was broken, releasing the mud that lay beneath. Even the Humber estuary – a mud bowl whose waters are now as murky as a hedge fund's tax returns – was once lined with oyster reefs. On the tidal slops, Callum Roberts tells us, you can still find oyster shells 'smoothed by more than a century of tides'. By creating, through the accumulation of cemented shells, a hard bottom onto which other oysters could attach, the shellfish, like cod and green turtles, engineered the environment that suited them. They also provided a substrate onto which many other species could attach, in turn creating habitats for yet more wildlife.

In Chesapeake Bay on the Atlantic coast of the United States there were sufficient oysters, according to one paper, to have 'filtered the equivalent of the entire water column every 3 days'.[32] As the early colonizers broke the land, much of the soil – and the nutrients it contained – began to wash into the sea. This process – called eutrophication – has been blamed for the periodic bloom of plant-like plankton, whose decay and nocturnal respiration sucks the oxygen out of the water, killing many of the animals the bay contains. This plankton contains species which poison the water, causing lethal red tides. Fascinatingly, however, despite the great dump of nutrients into the bay from around 1750 onwards, it was not until the 1930s, when the oysters had been more or less fished out, that such disasters began to occur.[33] The oysters filtered and consumed the plankton, preventing it from blooming and from poisoning the ecosystem. The damage, from the 1930s onwards, was self-perpetuating. As the oysters were reduced to the point at which they could no longer keep the water clear, they began to suffer from a lack of oxygen and the overabundance of sediments. This made them susceptible to disease, which further reduced their number. The report describing this effect remarks that Chesapeake Bay, the Baltic, Adriatic and parts of the Gulf of Mexico, are now 'bacterially dominated ecosystems'.[34]

The Black Sea also appears to have been transformed by the removal of some of its dominant species. After its predators – such as dolphins, bonito, mackerel and bluefish (*Pomatomus saltatrix*) – were reduced by commercial fishing, the plankton-eating fish they preyed on proliferated.

The result is that animal plankton numbers crashed, which meant that plant and plant-like plankton multiplied, sometimes poisoning the water, often depleting it of oxygen.[35] When the anchovies on which the predatory fish once preyed were then over-harvested, and comb jellies from the Atlantic arrived in the ballast water of ships in the 1980s and were able rapidly to occupy the depleted ecosystem, the chain of destruction came close to completion.

One of the most visible transformations has been the apparent shift from fish to jellyfish. The fishing trip I described in the second chapter was the last occasion on which I have taken even a moderate haul from the coast on which I live. I have launched my kayak dozens of times in the three years since then and not returned with more than two fish. This astonishes me in view of the abundance I encountered when I first arrived in Wales. Then, a single trip would supply as much fish as my family could eat in the season. On some occasions, in just a couple of hours, I caught as many as 150 mackerel, as well as weavers, gurnard, whiting, pollock, codling and scad. (I returned the rarer and smaller fish.) Those were thrilling moments: pulling up strings of fish amid whirling flocks of shearwaters, gannets pluming into the water beside my kayak, dolphins breaching and blowing. It was, or so it seemed, the most sustainable of all the easy means of harvesting animal protein. Now, for reasons I have not been able to identify clearly, that brief era – my first two years in Wales – has passed. I was surprised to discover that the fisheries officials and scientists I spoke to not only had no explanation for this apparent change; they had no data either. If there has been, as I suspect, a population crash, no one is studying it.

Something else appears to have changed. In the past two years Cardigan Bay has swarmed with jellyfish – not the little transparent moon jellies with which I was familiar, but species I had seen only rarely in the three previous years. Most of them are barrel jellies: solid rubbery brutes the size of footballs. Pale and ghastly, they fade the green depths; sometimes the sea appears to contain as much jelly as water. (I should emphasize that these are not scientific surveys; I am relating unquantified impressions. Unfortunately, in Cardigan Bay, there is no better source on which to draw.)

While the apparent transformation in Cardigan Bay has not been quantified, in the Irish Sea as a whole, and beginning long before I arrived on the Welsh coast, the ecosystem does appear to have been turning to jelly. A research paper links this change to a combination of warming waters and overfishing, in particular the herring fishery off the coast of Ireland in the 1970s.[36] There, fishermen using paired trawlers pursued juvenile herring to turn into fishmeal:[37] they were ground into feed for pigs and chickens or fertilizer for crops and lawns. I struggle to find the words required to describe the wasteful-ness of this operation.

This, the study suggests, might have helped to create 'a cascading regime shift', which tipped the balance in favour of jellyfish. With fewer competitors for the plankton they eat, they were able to prolif-erate. As the herring population begins to recover, this might go into reverse, though if the mackerel have gone, the jellyfish could once more have been released from competition.

Similar shifts have taken place, for the same reason, off the coasts of Namibia and Japan and in the Black, Caspian and Bering seas.[38] In all these cases, small plankton-eating fish, such as herring, sardines and anchovies, which both competed with the jellyfish for prey and, perhaps, ate the young jellies, have been greatly reduced by fishing, and animate gloop has swarmed into the breach. Jellyfish can also survive much better than fish in water whose oxygen has been depleted by plankton blooms: they are among the few lifeforms that can live in the dead zones now developing in many seas. They also have a pecu-liar ability to resist the destruction caused by fishing nets: they can regenerate themselves after they have been shredded.

One paper warns of a 'never-ending jellyfish joyride'.[39] Beyond a certain density, jellyfish inflict on depleted populations of herrings and similar species what the herrings inflict on depleted cod: they pre-vent them from recovering by eating their eggs and young. This allows the jellyfish to proliferate further, wiping out other fish and threaten-ing to replace them with a jelly monoculture.

The lesson emerging repeatedly from studies of the ecosystems of land and sea is that plagues take place when keystone species are removed. When they have not been heavily exploited, natural systems can, it seems, prevent explosions of native species and control inva-sions of most exotic species. They are also better able to withstand other disturbances, such as climate change, pollution, disease and storms. The planet was, before its foodwebs were broken up, controlled by animals and plants to a greater extent than most of us imagined. Evidence supporting James Lovelock's 'Gaia hypothesis' – that the earth functions as a coherent and self-regulating system – appears, at the ecosystem level, to be accumulating.

Our understanding of these issues suffers, like our perception of the state of the hills, from Shifting Baseline Syndrome. It applies throughout the ecosystems with which we engage, but it is especially powerful at sea, where fisheries scientists often recommend that stocks be restored to the state they recorded at the beginning of their careers, apparently unaware that this state was itself badly depleted. The past abundance described by explorers, naturalists and seafarers is often dismissed as fishermen's tall tales. On behalf of the peculiar tribe of anglers to which I belong, I feel obliged to admit that on a few, entirely unrepresentative occasions, we have been known to exaggerate. But the remarkable wealth of the seas before large-scale fishing began is also attested by more reliable evidence.

An article published in the journal *Nature* used government fisheries reports dating back to 1889 to estimate the extent to which fish populations in the North Sea have been depleted.[40] The results have revolutionized our understanding of the life it once supported. Instead of simply charting the amount of fish caught there, which creates the impression that the decline of fish populations has been moderate, it divided the fish caught by the amount of fishing power used to pursue them: the size and catching ability (larger engines, better nets, electronic fish finders) of the boats being launched.

When the British government first started gathering data, sail trawlers were beginning to be displaced by steam. Trawling in the North Sea had already been happening for 500 years, which means that the ecosystem was likely, by 1889, to have been gravely depleted. Even so, the researchers realized that, when fishing effort was taken into account, fish populations had declined not by 30 or 40 per cent in the following 118 years, as the scientists advising fishery managers had assumed, but by an average of 94 per cent. In other words, just one seventeenth of the volume of fish that existed in 1889 survived into the first decade of the twenty-first century. Fish stocks, they found, collapsed long before the amount of fish being landed declined: the landings were sustained only by ever more powerful boats, with ever more effective gear, scouring ever wider expanses of sea.

Haddock, they noted, had fallen to 1 per cent of their former volume, halibut to one-fifth of 1 per cent. But the most remarkable revelation in the paper was this: that in 1889 the fishing fleet, largely composed of sailing boats, using primitive, homespun gear, reliant on luck and skill rather than on fish-finding technology and all the other sophisticated equipment available today, landed twice the weight of fish as boats working the same sea do today.

Studies using different techniques have come to similar conclusions, both in our own seas and in other parts of the world: typically fish populations have been reduced by 90 per cent or more.[41] Yet so powerful is Shifting Baseline Syndrome that even some professional ecologists are snared by it. The UK's National Ecosystem Assessment, for example, which is generally a reliable guide to the state of the natural world, reports that 'around half ... UK finfish stocks [are] now at full reproductive capacity and harvested sustainably'.[42] Yet the baseline against which it makes this judgement is the state of stocks in 1970. By then they had been reduced to a small fraction of their 'full reproductive capacity'.

The same applies to the size of the fish that used to be caught, tales of which are frequently mistrusted by those suspicious-minded people who have never picked up a fishing rod. As the great fisheries scientist Ransom Myers found when surveying records of the first commercial fisheries on the ocean frontier, in twenty years the average weight of the tuna caught falls by half, while that of marlin falls by three-quarters.[43] There lived dragons where none live now.

Heavy exploitation began in many places long before the Industrial Age. The first known ecological complaint about destructive fishing techniques is contained in a petition submitted to Edward III, in 1376:

> the great and long iron of the wondryechaun runs so heavily and hardly over the ground when fishing that it destroys the flowers of the land below the water there, and also the spat of oysters, mussels and other fish upon which the great fish are accustomed to be fed and nourished.

By which instrument in many places the fishermen take such quantity of small fish that they do not know what to do with them; and that they feed and fat their pigs with them, to the great damage of the commons of the realm and the destruction of the fisheries.[44]

A wondryechaun is an object of amazement. The object in this case was a beam trawl pulled by a sailing boat. The flowers of the land below the water is an excellent description of the lifeforms – the soft corals, sea fans, sea pens, tube worms, fan mussels and all the other delicate creatures ('huge sponges of millennial growth and height . . . unnumbered and enormous polypi') – which must once have thronged the seafloor around our coasts but which are now rare or missing almost everywhere. And catching juvenile fish to feed to pigs? As the case of the Irish herring trawlers I mentioned a few pages ago suggests, not a lot changes.

The early industry sometimes managed to inflict great damage. The Scania herring of the western Baltic, for example, became extinct in the Middle Ages as a result of improved netting technologies.[45] Significant ecological change may go back even further. The excavations at Bouldnor Cliff, on the Isle of Wight (off the coast of southern England), for example, suggest that the Mesolithic people who lived there 8,100 years ago could have been running a boatyard. The wood-working techniques they used were previously believed to have arisen in Britain only 2,000 years later, in the Neolithic. Among the discoveries are a plank split from an oak trunk likely to have been used to make a log boat, and a platform that might have been used as a jetty or quay.[46] This suggests a fishing capacity greater and more sophisticated than previously imagined. Whenever a new fishery opens, the largest animals tend to be caught first. Who knows what monsters might have been extracted then? Ours is a dwarf and remnant fauna, and as its size and abundance decline, so do our expectations, imperceptibly eroding to match the limitations of the present.

It is not my purpose to dwell at length on the destructive habits of the fishing industry, some of which are likely to be well known to you. But I will briefly mention a handful, of which you might not be aware, which emphasize the need for a radical change in policy.

Every year the taxpayers of the European Union give €1.9 billion to the European trawler companies ransacking the fisheries of West Africa.[47] Once rich in a remarkable variety of species, the continental shelf there has been stripped by foreign boats, destroying the ecosystem as well as the livelihoods of local fisherfolk, whose boats and impacts are much smaller. Fish is an essential source of protein for communities in West Africa, but the foreign fishing fleets have wrecked many of the stocks on which they depend. One estimate suggests that the volume of unwanted fish discarded dead or dying by a single trawler on a single voyage in these waters is equivalent to the annual consumption of 34,000 people.[48] Ninety per cent of the licence fees the trawler companies would otherwise have paid to exploit these stocks is provided in the form of subsidies by the European Union and European governments. I wonder how many taxpayers believe that this is a good use of their money.

An investigation into a £63 million illegal fishing racket in Scotland discovered that a government body, Seafish (which 'supports all sectors of the seafood industry'), took a £434,000 cut.[49] Seafish is funded by a levy on the fish landed in the United Kingdom. It admits that it was aware that the Scottish fish were illegally caught, but, after consulting its lawyers, it continued to collect its fees. Chris Middleton of Seafish told me there was 'no need' to hand the money back to the government, and that 'there's been no call to do so'. Green campaigners claim that Seafish tries to undermine their efforts to prevent overfishing and that it defends destructive fishing practices against reform; the organization denies these charges. While other public bodies have been shut down or trimmed by the government, Seafish remains uncut and unreformed.

European fisheries help to supply Japan, whose government appears unmoved by the status of the species the country imports. Scarcity appears to stimulate its market. Charles Clover's film *The End of the Line* presented evidence suggesting that the electronics company Mitsubishi, which controls 40 per cent of the world market for bluefin tuna, has been stockpiling frozen carcasses, which can be sold at many times their current value when the species becomes commercially extinct. The company denies this.

When an international meeting in Doha tried to ban the trade in bluefin tuna – now as endangered as tigers and rhinoceroses – the Japanese government, much as it has done during negotiations over whaling, bought the votes of enough poorer nations to block the attempt. As if to underline its contempt for efforts to protect this magnificent animal, at a reception a few hours before the vote was taken the Japanese embassy served bluefin tuna sushi to its guests.[50] At the same meeting, Japan also managed to defeat attempts to regulate the international trade in corals and to protect some of the sharks that are hunted for their fins.

The demand for bluefin tuna, like that for rhino horn, shows no sign of declining as the fish becomes rarer. Rather, the fish is simply becoming more expensive. In 2012 a single bluefin was sold in Japan for £470,000.[51] The restaurant owner who bought it said he bid so high in order to 'liven up Japan'. He won the undying gratitude of his customers by selling cuts from the fish at below cost price.

We rightly deplore the apparent unconcern with which this species is being driven to extinction. But it is not a world apart from the habits of liberal, well-educated people I know in Britain – friends and relatives among them – who, despite widespread coverage of the impacts of unsustainable fishing on television and in the newspapers they read, continue to buy species such as swordfish, halibut and king prawns, which are either in dire trouble or whose exploitation causes great ecological damage.

To meet this demand, the world's continental shelves are being trawled, destroying their sessile lifeforms – the trees of the sea – at 150 times the rate at which forests on land are cleared.[52] In other words, every year half the global continental shelf is trawled. At this rate, it is impossible for the delicate animals destroyed when nets, beams, rakes and chains were first dragged over them to re-establish themselves. As farming and some varieties of conservation do on land, fishing reduces complex, three-dimensional habitats to featureless plains.

Until recently, much of the seabed was protected by the fact that it was rocky, and would damage any nets pulled over it. It provided a sanctuary for species extirpated elsewhere. But the rockhopper equipment developed in the 1980s and now used widely has made almost every hidden corner accessible. Those of us who enjoy exploring the shoreline are advised not to turn over rocks, for fear of crushing the creatures that live under them or on top of them, and depriving animals of their habitat. But across great tracts of sea, rockhopping trawlers turn over boulders of up to 25 tonnes,[53] either flushing out or smashing the fish and crustaceans they harbour, destroying the habitat as effectively as a bulldozer in a rainforest.[54]

Sometimes I wonder what hold the fishing industry – a small component of the European economy – has over ministers and members of parliament. Does it sink the bodies of their political opponents? Does it deliver the cocaine they use? While I doubt the reasons are as exotic as these (except perhaps in Italy), the political power of this industry is often mystifying. Perhaps the most likely explanation is that while many voters are upset by its destructive practices, few have as strong an interest in curbing them as the fishing companies have in perpetuating them.

It took hunters and farmers millennia to inflict as much damage on the life of the land as industral fishing has inflicted on the life of the sea in thirty years. But, if this feeding frenzy can be restrained, the restoration of marine ecology will be easier than restoring terrestrial ecosystems, for two reasons. The first is that few marine species of the continental shelves, even among the megafauna, have yet become universally extinct. (This is likely to contrast with animals living around the abyssal seamounts, many of which are found only in one place, are poorly documented and very slow-growing, and are now being heavily exploited by trawlers.) There are some well-known exceptions, such as Steller's sea cow and the Caribbean monk seal. But even animals which have been reduced to 1 per cent or less of their original populations – certain species of shark, tuna and turtle, for example – have, so far, clung on. There is enough time – just – to prevent them from disappearing for ever.

The second reason is that most of the species which live in the sea can reintroduce themselves to habitats from which they have been removed. Either the adults are very mobile (many fish and mammal species migrate hundreds or thousands of miles) or the eggs or young are released as plankton, which can drift great distances on the currents, like marine thistledown.

There is one sure means by which the ecology of the seas can be protected and restored. That is the creation of marine reserves in which no fishing or other industry takes place, and in which both mobile and sessile lifeforms are allowed to recover. In other words, rewilding.

In 2002, at two world summits, governments promised to protect at least 10 per cent of the world's seas by 2012.[55] In 2003 the World Parks Congress called for at least 20 or 30 per cent of every habitat at sea to become strict reserves by the same date.[56] Despite the creation of a few very large conservation areas, such as the Great Barrier Reef Marine Park, covering 350,000 square kilometres, at the time of writing less than 2 per cent of the world's seas has any form of protection,[57] and only in some of these places is fishing wholly excluded.

In 2004 the British government's official advisers, the Royal Commission on Environmental Pollution, proposed that 30 per cent of the United Kingdom's waters should become reserves in which no fishing or any other kind of extraction happened.[58] In 2009 an environmental coalition launched a petition for the same measure – strict protection for 30 per cent of UK seas – which gathered 500,000 signatures.[59] Yet, while some nations, including several that are much poorer than the United Kingdom, have started shutting fishing boats out of large parts of their seas, at the time of writing we have managed to protect a spectacular 0.01 per cent of our territorial waters: five of our 48,000 square kilometres. This takes the form of three pocket handkerchiefs: around Lundy Island in the Bristol Channel, Lamlash Bay on the Isle of Arran and Flamborough Head in Yorkshire. There are plenty of other nominally protected areas but they are no better defended from industrial fishing than our national parks are defended from farming.

When fishing stops, the results are remarkable. On average, in 124 marine reserves studied around the world, some of which have been in existence for only a few years, the total weight of animals and plants has quadrupled since they were established.[60] The size of the animals inhabiting them has also increased, and so has their diversity. In most cases the shift is visible within two to five years.[61] As the slower-growing species also begin to recover, as sedentary lifeforms grow back and as reefs of coral and shellfish re-establish themselves – restoring the structural diversity of the seabed – the mass and wealth of the ecosystem is likely to keep rising for a long time.

Five years after Georges Bank, off the coast of New England, was closed to most forms of commercial fishing, the number of scallops had risen fourteenfold. Around Lundy Island, mature lobsters trebled in number within eighteen months of the creation of the reserve.[62] After four years they were five times as abundant as those outside;[63] after five years, six times.[64] Eighteen years after they were first protected, the combined weight of large predatory fish in the Apo Island reserve in the Philippines had risen by a factor of seventeen.[65] Bigger fish produce more eggs, and the quality of the eggs improves as the parents mature, so more of the offspring are likely to survive. Like the Kraken in Tennyson's poem, the suppressed life of the sea awaits only its chance to re-emerge.

Not all missing populations can be restored. Some of the lifeforms being wrecked by perhaps the most destructive fishing operation of all – the trawling of the deep seamounts – take thousands of years to grow. Many of them are endemic, confined to one place. Extinction there is extinction everywhere. Scientists are also beginning to understand the extent to which some populations of fish are specific to particular spawning grounds. Like salmon returning to the rivers of their birth, for example, every population of cod appears to possess its own migration routes, and travels to particular banks and reefs to reproduce, following invisible rivers beneath the sea. This could offer another explanation for the failure of some cod populations to recover after fishing for them has ceased: if one group has been destroyed, neighbouring communities are unlikely to fill the gap, just as salmon born in the River Tweed will not replace the salmon missing from the Thames. Migrations are led by the bigger, older fish, which are the first to be exterminated by overfishing.[66]

Nor would it be correct to suggest that reserves are the only necessary measure. There should also be restrictions on the kind of equipment used in places where fishing continues, on the capacity of fishing boats and the time they spend at sea and on their freedom to discard the fish they do not want to keep. The reserves will probably work best if they are surrounded by zones in which pressure is reduced: where, for example, only line fishing is allowed. But the rewilding of parts of the sea is the essential element without which protection is almost meaningless.

There are fewer inherent conflicts between marine rewilding and those who make their living from the seas than there are between terrestrial rewilding and those who make their living from the land. Biologists have noticed a strong spillover effect: the fisheries surrounding marine reserves improve because the spawning fish are protected and allowed to reach maturity, and they and their offspring migrate into surrounding waters.

Fishermen tend to resist marine reserves before they are created, then to support them once they have been established, as their catches rise, often far beyond expectations. In the seas surrounding the Apo reserve that I mentioned a moment ago, for example, the catch swiftly rose to ten times its previous level, and has stayed that way since.[67] There have been similar results in, for example, fisheries off Japan, New Zealand, Newfoundland and Kenya.[68]

Marine protection is so cheap and the results so lucrative that, the Royal Commission calculated, just a 2 or 3 per cent increase in the fish catch in the North Sea would pay for the protection of 30 per cent of its area.[69] The returns are more likely to rise by 200 or 300 per cent.

A report by the New Economics Foundation suggests that the failure to protect fish stocks properly costs the European Union some 82,000 jobs and €3 billion a year.[70] Marine rewilding not only offers the best chance of protecting much of the life of the seas, but also the best chance of protecting the livelihoods of those who harvest it. The weight of fish landed worldwide peaked in 1988. Despite attempts by Chinese officials to inflate their production figures, it has been declining since then by half a million tonnes a year.[71] The surest means by which this could be reversed is the creation of a network of large marine reserves.

But here, as so often, we see short-termism triumph over not only wider social and environmental interests, but also over the medium- and long-term interests of the people who block this reform. For example, the proposal to stop crab and lobster fishing in just 1,100 hectares around Skomer Island, off the coast of Pembrokeshire in Wales, was voted down by the fishermen on the committee which considered it,[72] despite the evidence of greatly improved catches around similar reserves. The prospect of lower returns for the first one or two years of the reserve's existence appears to have outweighed the promise of higher returns for ever after.

The opposition of the fishing industry also explains the dithering and downgrading by the British governments which promised to protect the life of the seas. In 2004 the Royal Commission pointed out that the seas around this country 'have been scrutinized in great detail since at least the mid-19th Century'. Existing data was easily sufficient 'to design comprehensive, representative and adequate networks of marine protected areas for UK waters'. But at the time of writing, eight years later, the Westminster government is still procrastinating, on the grounds that 'there are a number of gaps and limitations in the scientific evidence base'.[73]

The government originally offered to protect 127 sites in English waters. Now it appears to be paring the list down. Worse still, it intends to protect only the remaining 'vulnerable features'. In most places trawling has already destroyed just about every fragile habitat; the government, according to a conservationist heavily involved in this debate, intends to 'protect the pin-pricks of features that remain, and allow trawling around them . . . Someone recently likened this to designating a ploughed field for an oak tree in the middle of the field, and only the oak tree is protected, whilst the ploughing is continued.'[74]

Even this feeble protection will apply to only some of the sites on the list: the government says that designating an area as a marine conservation zone 'does not automatically mean that fishing in that site will be restricted'.[75] Many of them will be protected in name only. Unless something changes, the reforms will raise the proportion of England's seas in reserves where no fishing takes place to around 0.5 per cent: one sixtieth of the level the Royal Commission suggested was necessary to protect a significant portion of marine wildlife.

In Wales the policy is even worse. The government has promised to consider 'no more than 3 to 4 sites',[76] covering 0.15 per cent of its seas.[77] So far there has been no certain progress towards this miserable target. The 'protected areas' we already possess are nothing of the kind. For example, a little way down the coast from where I live is the Cardigan Bay Special Area of Conservation (SAC). Special areas of conservation are supposed to offer the highest level of protection available under European law. They are described by the government's official conservation body as 'strictly protected sites'.[78] Yet in the Cardigan Bay SAC, set aside, we are told, to protect Europe's largest population of bottlenose dolphins[79] and the rest of the life that persists there, every form of commercial fishing bar one is unrestricted, except by the laws that apply to unprotected areas. The only commitments in the management plan are to 'review' and 'assess' the fishing that takes place, to 'encourage' good fishing practices (without discouraging the bad ones), and to ask fishermen to record the dolphins and porpoises they accidentally catch and kill.[80] That, dear reader, is 'strict protection'.

As a result, beam trawling, otter trawling and, with one exception, any other forms of industrial fishing the boats wish to pursue continue there unhindered. There is no prospect of the seabed or the ecosystem recovering from past destruction as a result of this regime of 'strict protection'. Nor is there an opportunity for the fish stocks on which the dolphins depend to rebound.

There is one method – scallop dredging – that is restricted in some parts of the Special Area of Conservation. With the possible exception of dynamite fishing, it would be hard to devise a more effective means of destroying both living creatures and their habitats. Scallop dredges operate by raking through the seabed with long metal teeth, dislodging the shellfish from the sediments and trapping them in a net whose underside is made of chain mail. The teeth rip through any sedentary creature in their path, as well as the fish, crabs and lobsters unable to escape in time. The steel mesh smashes animals missed by the teeth. Where they are used, divers publish heartbreaking photographs of the seabed before and after they have passed. It looks, where the dredges have worked, like a ploughed field, lifeless, covered in fragments of shell.

As if to demonstrate what 'strict protection' really means, the Welsh government decided to let the scallop dredgers into the middle of the reserve. The government's official advisers, the Countryside Council for Wales, warned that if dredging went ahead it would be 'likely to have a significant effect on Cardigan Bay SAC' and may have 'adverse effects on the dolphin population'.[81] This advice was ignored, and dredging was permitted within a large square at the heart of the reserve. This also served as an open invitation for these scarcely monitored boats to leak into the surrounding waters and dredge the other parts of the SAC.

The person responsible for this decision was Elin Jones, the minister then in charge of rural affairs with whom I had the frustrating exchange over farming policy. After I had asked her about the management of the uplands, I turned to scalloping. She told me she was 'not convinced that it has any kind of degrading effect on the SAC'. She agreed that the Countryside Council for Wales had warned against it, but said she had taken advice from another body, called CEFAS (The Centre for Environment, Fisheries and Aquaculture Science). This agency lists among its duties 'collaborations with the fishing industry'[82] and 'address[ing] fishermen's own concerns about scientific assessments'.[83] Why, I asked, did she take this body's advice while rejecting the Countryside Council's?

'Well, because I was more convinced by the CEFAS advice than by the CCW advice.'

'What was it that convinced you?'

'Well, I can't recall at this particular point on that.'

I pressed her further on the reasons for her decision. She told me she wanted to strike a balance between 'the need to protect our seas' and 'to protect and even enhance the coastal fishery that we have'. So, I asked, given that most of the dredgers come from Scotland and the Isle of Man, fish scallops for a few weeks and then move on, how do their activities enhance the Welsh coastal fishery?

'Well, it means that people in Aberystwyth and Machynlleth who want to eat scallops can hopefully eat scallops from Cardigan Bay ... scallops are eaten by people from this area, and I want them to be fished from as close a source as possible.'

I pointed out that the great majority of the scallops taken from the bay are shipped abroad, mostly to Spain, France and other parts of Europe.

'Yes, I know, that's part of the weakness of what we have at the moment, which is something that I'm trying to address, through the funding activities that we have in the European Fisheries Fund, to improve the quayside infrastructure that we have in Aberystwyth or

Cardigan, to ensure that all of this fish that's caught can be kept locally and sold locally.'

A local fisherman tells me that some £6 million worth of scallops is caught in Cardigan Bay every year. The population on and around the bay is tiny and overwhelmingly poor. The dredging industry exists because of lucrative markets abroad. There is no obvious mechanism by which local people could outspend these markets, even if they developed a sudden craving for *coquille St Jacques* at breakfast, lunch and tea. In other words, of the many unlikely propositions I have heard issuing from the mouths of ministers, this must be the most ridiculous.

One October, two years after I discovered that lime trees were growing there, I returned to the Nantgobaith gorge with a friend. Instead of following the forestry track on the north side of the little river, on which I had found the leaves which suggested that this might be a fragment of primeval rainforest, we slithered down the steep south bank of the gorge. We wanted to walk where no one had walked for many years, and to see which trees were growing in the scarcely accessible parts of the wood.

This fragment must owe its existence to the topography: the land is, or was, too steep to clear and too dangerous for keeping sheep. We slipped and slithered in the soft black loam which barely coated the rocks to which the trees clung. Below us the river roared through narrow passages and over cataracts. Had we lost our footing we could have slid down the gorge to our deaths. Clinging by our fingers to exposed roots, the stems of saplings, slippery emergent rocks, we slowly lowered ourselves towards the valley bottom.

When we reached the river, we began picking our way over moss-slick boulders in the mist raised by the many rapids and falls. Before long we came to a white chute of water between two crags. Standing on one of them, I peered gingerly over the edge. 'Wouldn't it be amazing,' I asked, 'if we saw a salmon leaping the falls?'

'I would love to see that.'

'I doubt they run up this river. And it's probably the wrong time of – bloody Norah!'

As if I had summoned it, something bronze and glistening arced out of the water, failed to reach the top of the falls and crashed back into the plunge pool.

'Did you see that?'

'No, what?'

'I saw a salmon.'

'You're kidding.'

'Watch.'

A minute later another fish hurled itself into the air. We sat on the rock and brought out our lunch, and for the next hour or so we watched salmon large and small rising from the water, wriggling through the air as if to gain purchase on it then tumbling back into the white chaos from which they had sprung.

Willing them upwards, elated by their flight, catching my breath every time a fish appeared, I found myself enraptured. I felt at that moment as if I had passed through the invisible wall that separated me from the ecosystem, as if I were no longer a visitor to that place but an inhabitant – a bear perhaps, emerged from a bimillennial absence back into this ancient scrap of wildwood (which might indeed have been one of the last places in which its species held out), leaning over the falls, mouth agape, fur sodden with spray, knowing at that moment only the water and the fish and the rocks on which it stood.

It was then that I realized that a rewilding, for me, had already begun. By seeking out the pockets of land and water that might inspire and guide an attempt to revive the natural world, I had revived my own life. Long before my dreams of restoration had been realized, the untamed spirit I had sought to invoke had already returned. By equipping myself with knowledge of the past while imagining a rawer and richer future, I had banished my ecological boredom. The world had become alive with meaning, alive with possibility. The trees now bore the marks of elephants; their survival in the gorge prefigured the return of wolves. Nothing was as it had been before. Like the salmon, improbably returning from the void, the depleted land and sea were now gravid with promise. For the first time in years, I felt that I belonged to the world. I knew that wherever life now took me, however bleak the places in which I found myself might seem, that feeling – the sense of possibility and, through possibility, the sense of belonging – would remain with me. I had found hope where hope had seemed absent.

Salmon are not the only fish which, in some parts, are beginning to recover. While the bottom-dwelling creatures of the sea cannot rebound without the closure of fishing grounds, some of the pelagic animals – those unattached souls which haunt the middle reaches of the sea – have begun, in a few places, to demonstrate the remarkable capacity of marine life to reappear when conditions change. The cessation of the fishmeal operations in the Irish Sea and elsewhere, and the gradual, partial recovery of the herring population, has attracted back into our waters animals which were once common here.

In 2009 a lone killer whale joined the dolphins that reside in Cardigan Bay. At one point it turned up half a mile off the beach at Llansglodion. A small pod – always accompanied by the same large male – has now appeared off the coast of Pembrokeshire every May for the past eight years. Minke and fin whales have arrived in the same seas in numbers unknown for decades. In 2011 the fins – the second-largest animals ever to have darkened the planet – began, for the first time in living memory, disporting themselves off Pembrokeshire in the winter as well as the summer.[84] Later that year, twenty-one of them were seen in the Celtic Deep.[85] In 2005 a humpback whale was spotted off the seaboard of Wales. Two others arrived in the Irish Sea in 2010; one was filmed breaching three miles off the Irish coast.[86] I dream of the day on which I kayak among them.

From the North Sea come, once more, reports of gigantic fish hammering the mackerel shoals off the Yorkshire coast, and occasionally stripping all the line off an unsuspecting angler's reel: though the bluefin tuna is critically endangered, some among the depleted population are again following their prey into these waters.[87] A few years ago a much commoner tunny, the albacore or longfin, began harrying the herring off the coast of Ireland. In three consecutive years, fishermen had reported shoals of albacore, some leaping clear of the water, less than a mile from the coast on which I live. It was this latter intelligence which led me into the stupidest and most dangerous adventure upon which I had embarked for, oh, at least a month.

14

The Gifts of the Sea

Many times have I stolen gems from the depths
And presented them to my beloved shore,
He takes in silence but still I give
For he welcomes me ever.

> Khalil Gibran
> *Song of the Wave*

Though I sought to persuade myself otherwise, in my heart I knew that I had no hope of finding or catching an albacore. I later discovered that a kayak cannot travel fast enough to pull a lure through the water at the necessary speed. I suspect and hope that, had I not been half-aware of the futility of my quest, I would not have embarked upon it. I had no desire to kill such a creature, or to inflict pain on an animal I did not intend to eat. Nor had I any idea of what I might do if by some extraordinary fluke I managed to hook one. But the thought of it – the dream of it – pulled me away from my desk, on a buoyant, glittering day in early October.

The river, swollen by a summer of incessant rain, roared down to the sea. The water fountained into the air where it hit the first rocks. Below them it furrowed into foaming gullies and wild rides, swerved against the banks, whirled round in an exultation of flying spray then exploded once more on contact with the next set of rocks. It was, in a thirteen-foot sea kayak, an interesting passage.

I hit the waves in the river's mouth with a smack. A stiff westerly had stacked the breakers against the shore, twenty or thirty deep. I dug in and fought what seemed to be a losing battle to break through them. I began to suspect that I was not progressing at all, but only sliding forwards on the backwash then backwards on the incoming wave, again and again. But at last I burst through the back of the surf and out into the most exhilarating sea I have ever sat upon.

It was a magnificent mess. The south-westerly swell mounted and tumbled against the west wind. No wave resembled its predecessor. Sometimes the peaks and troughs cancelled each other, and I found myself marooned on a raft of flat water. At other times they coalesced. The sea would suddenly give way beneath me and suck me into a square-sided hole, or two or three waves would join forces and lift me high into the air until my kayak teetered on the edge of a chalcedony cliff before free-falling into the gully behind it, landing with a great jolt and an explosion of spray. White horses reared up from nowhere and came down upon my shoulders with a clatter of hooves.

The forecast had told me that the wind would drop towards evening, but now it was lively and thrilling. The sun capered across the waves, its sport threatened by nothing but a faint smoke of high cirrus and a few puffy cumulus low on the horizon. I paddled out far enough to ensure that I would not be blown back into the breakers while I rigged the tackle. As soon as I stopped moving, the boat swerved and tilted, threatening, as it swung broadside, to tip me into the water with every wave that passed beneath it. Gingerly, aware that if I let go for an instant I would lose irretrievably whatever I was holding, wobbling as I sought to keep my balance, I unpacked my stoutest fishing rod, and a new reel, loaded with hundreds of yards of line, that I had bought for this expedition. Trapping the paddle beneath my feet, I tied on a swivel and a rubbery artificial squid, masking a large hook. It looked ridiculous, like a toy children use to frighten each other. In my fishing bag, lashed with braided cord to a rear cleat, was my spare tackle, a water bottle, sandwiches and my waterproof camera: if the event I doubted, dreamed of and dreaded in equal measure were to occur, no one would otherwise believe me. I sunk the butt of the rod into the well behind my seat and, relieved still to be attached to the boat, set off.

My plan was to travel away from land towards the north-west until I was two miles from the shore, then to swing south, trolling the lure first in an arc against the swell, then parallel to the coast for a few miles, before I paddled back to the river's mouth. I had been told by more than one old salt that the fish were migrating and probably not feeding. I knew that my chances of attracting an albacore were minimal, and that, if such a miracle occurred, the question of who had caught whom would not be easily resolved. But the wild dream of it, goaded by stories that echoed in the catacombs of childhood memory and a yen for improbable glory, had dragged me, almost beyond will, into this furious sea.

Had the sun not been shining, had the sky and waves been cut from slate, not crystal, the sea that now looked inviting to me would have appeared forbidding, perhaps terrifying. But we are simple creatures, and a sprinkling of stardust dazzles our senses.

Had the quest not been so arousing, had the ride been any less thrilling, I might have challenged the grounds on which I continued to head out to sea. This is a roundabout way of telling you that I failed to notice that a journey which had begun foolishly was now progressing towards madness: the wind had both freshened and swung to the south-west. By the time I looked around to see what progress I had made, I had been blown two miles up the coast.

I decided to start fishing, sooner than I had intended, while paddling back along the shore. I paid out the line from the heavy reel, and the ridiculous creature fluttered away out of sight into the green water. I leant forwards and hacked at the sea, fighting through the waves, loving the sensation of the water streaming past the bows. But when I turned to check the marks, I realized that I had gone nowhere. Only then did I see how much trouble I was in. I stowed away the tackle as quickly as I could, strapping the rod to the boat once more, stuffing the tackle back into the bag and buckling it down securely. Even so, by the time I had finished I had drifted a good distance further up the coast. The wind had stiffened again and now it was coming from the south – directly against my line of travel. The shingle beach closer to the rivermouth had given way to rocks. To the north – the direction in which the wind was trying to push me – were cliffs. The tide was up and the south-westerly swell hammered into them. The breakers sounded like a motorway.

I put my head down and took on the wind. A kayak is a wonderful vessel; it can make way through remarkably high seas – as long as the wind is low. Against the wind it is a feeble instrument. There is a point – roughly eighteen knots, or a force five – beyond which it can make no headway: the resistance offered by the paddle and the body of the paddler matches his propulsive power. I managed, with a great expenditure of effort, to progress a quarter of a mile homeward, but then the wind rose once more. Now it whipped the crests off the waves, which came at me from all angles, barging me from one wall of water to another. The white horses ramped and whinnied, bucked when I vaulted onto their backs and lashed out with their hind legs as they passed. I rode these mustangs for another half-hour, during which time, to judge by the marks on the shore, I managed to cover fifty yards. Then I stopped. It felt as if someone had attached a tow rope to the back of the boat. However hard I paddled I could make no headway: in fact I seemed slowly to be travelling backwards.

I reviewed my options. If I gave up and stopped paddling I would be driven onto the cliffs. If I abandoned ship and swam I would, being some three-quarters of a mile from the shore, certainly lose the boat, possibly run out of energy and perhaps be dashed on the rocks when I arrived, though at least I would then have a chance of slipping under the breaking waves, rather than perching more perilously on top of them if I sought to beach the kayak. But landing in any condition among those rocks did not appeal to me.

Two hundred yards ahead of me I could see a small crescent of sand not yet covered by the rising tide; otherwise the beach gleamed with round boulders. The angle of attack was steep enough to negotiate, as I would be able to cut across the wind, but shallow enough to be dangerous when I reached the shore: I like to keep the waves directly behind me when I land, as that gives me a chance of controlling the boat. But I did not possess a surfeit of choice.

If I misjudged the angle or if I were blown back too far, I would miss the sand and find myself on the rocks. But the judgement was hard to make as the waves were so jumbled. I slid down their faces, fishtailing, tumbling, lurching towards the shore at alarming speed. Within a few minutes I found myself approaching the near horn of the crescent of sand. I was coming in tight and was in danger of missing it. I grunted with effort, trying to drive the boat further along the shore, feeling every muscle cell strain and twang. Then I turned the boat inland to try to ride onto the utmost corner of sand and heard, as I did so, a shocking sound.

I turned. The biggest wave I had seen that day, that year, was bearing down on me. It was a wall of brown water, dirty, riotous, fanged with spume and shingle. As it loomed over me it shut out the low sun. I had seen footage of waves like this, on whose faces tiny surfers, black and slim as water skaters, weave and bob, and I had marvelled at their courage or foolishness. And now –

The breaker lifted me until I looked down upon the rocks of the beach as if addressing them from a balcony. The nose of the boat tipped down, my stomach seemed to fall away, and the water rushed me forward at astonishing speed. I leant back on the kayak, eyes wide with fear. There was nothing I could do. If I used the paddle I would be more likely to capsize the boat than steady it. Even in the thick of terror, it was magnificent. For a second, fear and thrill mixed in equal measure. Then I saw where the wave was taking me, and the thrill died, snuffed like a candle. The wave had swept me past the sand and was driving me onto the rocks. I was about to leap from the boat when it corked from under me.

The breaking wave rolled me over, tipping me out on the landward side of the kayak. I squatted on the seabed, beneath the water, and shoved the boat up as it came down on top of me, threatening to crush my head. The next wave lifted it and smashed it with hideous force onto a boulder. The hollow boom it made resounded off the cliffs behind the beach.

I emerged from under that wave and found the shore beneath my feet. When I stood I discovered that the water was only waist-deep. I caught the paddle and waded onto the sand. The boat was sucked out again momentarily, then crash-landed once more, wedging among the rocks. I turned it over then pulled it higher up the beach. My rods were still strapped to the gunwhale and, surprisingly, unharmed. But my tackle bag had gone. The collision with the boulder had snapped the cord. In that wild water, with the tide still rising, there was no hope of finding it. I resolved to return at low tide the next day, but I knew that in a sea like this, with longshore drift doubled by a following wind, my chances of finding it were probably lower than my chances of catching an albacore had been.

I stared at the sea, cursing myself. I had thought I had grown out of this kind of idiocy. I could scarcely believe that I had lured myself into such danger. I thought of the duty I owed to my daughter and my partner. Though I had just emerged from cold water, I burnt with shame. I began to realize, too, that I was not yet out of it.

The beach on which I had landed is one of the least visited places on the Welsh coast. It was, as it almost always is at that season, deserted. The nearest road was far away. It was bounded by a low but unscalable cliff of glacial till: the slippery clay and round boulders dumped by the ice sheets. Between the cliff and the sea lay just three or four yards of strand, and the tide, my watch told me, was still rising. The beach was a maze of rocks, grey and tan, that had been eroded out of the cliffs. I looked along the shore, hazy with the spray kicked up by the waves, as far as I could, and felt overwhelmed by loneliness.

It was, I found, impossible to drag the boat over the exposed boulders, and on that lumpy beach I could not carry it more than a few yards. My only option was to pull it through the surf. I soon discovered how difficult this was. I could pull the kayak forward for a second or two on the incoming wave, but then it would tip over, sweep round and knock my legs from under me. On the outgoing wave it would advance a little, then suddenly become grounded and wedged between the boulders. Only as the backwash began could I make significant progress. So I waited, tried to hold it steady on the incoming wave, sprinted forward through a momentary patch of smooth water, then stopped dead as it thumped back down onto the beach.

Already sapped when I landed on the beach, I was becoming even more exhausted. Whenever the boat turned over, more water seeped through the hatch. As it became heavier, it became more dangerous. But to empty it I had to drag it out of the waves and over the boulders, which was also tiring. I was beginning to wonder how I would get back without abandoning my most precious material possession when an extraordinary thing happened.

I knew what it was the moment I saw it, but I refused to believe it. It was so improbable that I imagined for a moment, in my wretched state, that I was hallucinating. It simply could not be true. It was like seeing a zebra trying to hide among the dresses in a department store. But though I had never seen one before, and though, when I first spotted it, it was sixty or seventy yards away, I knew that there was nothing else it could be.

It walked with odd jerky movements, extending and retracting its long neck. It had a sharp little head, no tail, long, pale pink legs. It looked like a pullet on stilts. I tried to persuade myself that it was a water rail or even a partridge. But it was not. It did not try to fly away. Instead it made little panicky darts between the boulders, slipping and scrambling, one moment trying to run away from me along the beach, the next trying to scurry up the cliff, but slipping down, wings flapping furiously. I drew level with it and stood in the surf while the bird tried to flutter up the boulder clay, perhaps seven or eight yards away. I could doubt it no longer. I saw the chestnut flash on the wings, the sharp chicken's beak, the low slim head, the beautifully netted plumage on the back – black and buff – feathers ruffled by the wind as it turned and tried to slither away up the beach. It was a corncrake.

Though common in other parts of Europe, it is exceedingly rare in Britain, and has not lived in Wales for many years. There is a farmer, now well into his eighties, living in the Desert a few miles inland from my home, who recalls hearing them in his youth, but they have not bred in these parts since that period. Their decline, throughout Britain and Ireland, was, from the 1970s until the 1990s, precipitous,[1] though, with the help of conservation programmes, they are slowly beginning to recover.[2] The nearest populations to that lonely stretch of coastline in mid-Wales are in western Scotland and (though fewer still in number) in northern Ireland.

At first it just seemed wrong. That this delicate creature should pitch up on a grey, boulder-strewn beach, so far from home – it was as if nature had fused, short-circuited: 'A falcon, tow'ring in her pride of place, / Was by a mousing owl hawk'd at and kill'd.'[3] But then an explanation occurred to me. It must have been migrating south, following the coast, when it ran into the wind that had foiled me and was grounded, exhausted on the beach. Perhaps it had read the same misleading forecast as I had. As understanding dawned, so did the thrill of what I had witnessed. I felt, too, a sense of solidarity with this frail little bird, battling the same forces as me, trapped on the same diminishing strip of beach.

In pursuit of one returning animal I had encountered another. And this encounter was just as gratifying, just as enchanting, as contact with an albacore would have been. I had set out to find an early result of a fractional rewilding and, despite everything, had found it. Had it not been for the near-disaster from which I had just emerged, I would not have done so.

As the bird receded up the beach I felt my energy surging back. Buoyed, ecstatic, I fairly marched the next mile, crashing through the rocks and surf. Then the beach widened and I was able to drag the boat along a strip of smooth pebbles above the tideline. Within an hour of seeing the corncrake I came to the bank of the river, now dammed by the risen tide. I plunged in, dragging the boat behind me, but soon found myself out of my depth, so I swam across, towing the kayak. I reached the pebble beach on the far side. Beyond it were the long low slacks across which I could carry the boat back to the car.

I sat on the kayak, exhausted, watching the yellow sun arcing down towards the water. In the salt mist above the breakers gulls skated and jinked on the wind. The waves opened and closed their jaws, slick with sunlight. I felt a curious mixture of shame and triumph. I had confronted the casual power of nature and – no, not won, no one ever wins – survived.

At lunchtime the next day I drove through the long glacial valleys of Snowdonia. The trees and bracken had suddenly turned: the dull greens of late summer had burst, almost overnight, into russet and umber, ochre and flame. I travelled to the point at which the road came closest, a mile and a half to the north of the beach where I had lost the bag.

It was another bright day. The wind still blew strongly from the south (I thought of the poor corncrake) and the waves roared, now, at low tide, far from where I stood. I walked down the concrete steps from the campsite in which I had parked and stared at the beach. I was confronted by the impossibility of what I had set out to do.

The bag could have been anywhere along that coast. It might have reached Porthmadog by now, or been swept out to sea, or buried in sand or weed. Even if it were somewhere on the mile and a half of strand between where I stood and where I had landed, only a search party of hundreds would have stood a good chance of finding it. The beach at low tide was a quarter of a mile wide. Below me was a long sweep of sand from which grey rocks emerged. Closer to the water were craggy boulders, rockpools and deep beds of wrack and furbelows.

But I had used almost a litre of fuel to get here, it was low water slack and I was not going to give up yet. I crossed a small stream and stepped down onto the beach. The fragile sunlight lay on the sand like gold leaf. Over the water the haze of salt appeared to light up the sky. The rocks glistened, black, against the bright sea. I would enjoy the walk and search as far as I could before the tide shut the door.

I walked perhaps ten yards, then stopped. I had seen something blue sticking an inch or two from the sand. I stared at it stupidly for a moment. It looked like the top of my water bottle. I stared for a moment longer, as the cogs slowly clunked together and began to turn. It *was* the top of my water bottle. Around it, scarcely emerging from the sand, was a black lip of some kind. The basal ganglia registered the sight, but it seemed to take an age before the message bubbled up to the vaguely conscious sections of my brain. It was the flap of my fishing bag, wrapped around the bottle.

The chances of finding the bag were tiny. The chances of finding it within thirty seconds of stepping onto the beach were ... 1 in 10,000? 100,000? 1 million? I dug round it like a dog and heaved it out. It was rammed full of sand, but the clips were still closed. It must have weighed half a hundredweight. I blinked at it, then I hauled it onto my back and staggered away. Water poured from the bag and down my legs.

At home I filled a galvanized dustbin with water and emptied the bag into it, then felt around in the sand – gingerly as I was mindful of the hooks – with the thrill of a child plunging his hand into a lucky dip. I began pulling out my belongings: first a reel, then a tangle of lures and line from which my camera dangled, then the other reel, then the smaller tackle. Everything was there.

The reels and the camera were seized up with sand. Over the next few days I dissected all three. Mending the reels was not too difficult, but the camera appeared to be dead. I shook half a handful of sand out of it, dried off the parts then reassembled it. There was no spark of life. I do not like throwing things away, so I left it on a shelf. Two weeks later, without a thought in my head, I picked it up and pressed the power switch. It flashed on then off again. I recharged the battery and tried again, with the same result. After another week, it came on for thirty seconds before shutting down again. Over the following two months it slowly revived, regaining another function every time I turned it on. By Christmas it was working perfectly.

15
Last Light

The Mind, that Ocean where each kind
Does streight its own resemblance find;
Yet it creates, transcending these,
Far other Worlds, and other Seas;
Annihilating all that's made
To a green Thought in a green Shade.

Andrew Marvell
The Garden

I have one more thing to relate, and it is a small one. A few days after
the albacore hunt, I finished work early and took my boat down to
the sea for the last time that year. I had decided that I would leave
Wales. Though the reasons were happy ones, the decision was edged
with sadness.

It was a calmer day, although there was a long swell over the reef. I
thumped through the waves, following distant gannets which dispersed long before I reached them, travelling a couple of miles out to
sea before allowing the north wind to push me down the coast. After
two hours without fish, I began to plod back, against both wind and
waves. Were it not for the shipping buoy moving steadily past the distant houses as I paddled, I could have imagined I was merely stirring
the viscous water. Then, as the sun began to sink, the wind dropped.
At first the sea looked like the broken bottoms of wine bottles, each
wave a conchoidal fracture. Soon, but for a slight residual swell, it fell
flat. Now the boat, as if untethered, cut cleanly through the still water.

A few yards from the shore I wound up my line then sat without
paddling, rocked by the incoming waves, watching the sun go down

over Yr Eifl, many miles across the sea on Pen Lleyn. The mountain appeared to snag the star then to drag it down into the earth like an ant lion. A puff of indigo cloud, like cannon smoke, hung against a sheet of flaming cirrus.

I looked around the bay. Though the light was fading, I could see the whole crescent. To the south was the gently rising plateau of the Cambrian Desert, dissolving into the suggestion of Pembrokeshire, where a few lights now glimmered. Closer to where I sat, the yellow flanks of Cadair Idris still faintly glowed, richer in colour than they had been a moment ago. To the north were the peaks of Snowdonia, washed and blue at first then hardening as they swept towards the point at which the sun had set. The mountains of Pen Lleyn now towered out of the sea, every knot and cleft sharp against the dying light. Beyond them Ynys Enlli, whale-backed, rode the still water.

I thought of the places I would be leaving, of what they were and what they could become. I pictured trees returning to the bare slopes, fish and whales returning to the bay. I thought of what my children and grandchildren might find here, and of how those who worked the land and sea might prosper if this wild vision were to be realized. I thought of how, across these five years, my exploration of nature's capacity to regenerate itself, of the potential for wildlife to return to the places from which it had been purged, had enriched my own life. Wherever I went, I would take the wild life with me. I would devote much of my life to seeking out or helping to create places where I could hear again that high exhilarating note to which I had for so long been deaf, where I could find that rare and precious substance, hope. The black silhouettes of redshank and oystercatchers piped home along the shore. To the south, moonlight glittered on the water, now grooved like a linocut.

From behind me came a noise like a boot being pulled out of the mud. I turned, but all I saw was a large round ripple, as if a monstrous trout had sucked down a fly. Then a fin rose from the lavender sea, five or ten yards away. It sank again then rose beside me. It was a baby: one of last year's dolphin calves. It circled the boat, so close that it almost nudged my paddle, then disappeared into the darkness.

Notes

INTRODUCTION

1. Fred Pearce, 16 September 1996, 'The Grand Banks: Where Have All the Cod Gone?', *New Scientist*.
2. Lori Waters, 11 January 2013, 'Enbridge deleted 1000 km2+ of Douglas Channel Islands from route animations', http://watersbiomedical.com/islands/jrp.html
3. Carol Linnitt, 13 April 2012, 'Oil and Gas Industry Refused to Protect Caribou Habitat, Pushed for Wolf Cull Instead', http://www.desmogblog.com/oil-and-gas-industry-refused-protect-caribou-habitat-pushed-wolf-cull-instead
4. Nature News, 29 June 2011, 'Scat evidence exonerates wolves', *Nature*, vol. 474, p. 545, doi:10.1038/474545d, http://www.nature.com/nature/journal/v474/n7353/full/474545d.html
5. Maggie Paquet, 2009, 'Saving Caribou in BC', *Watershed Sentinel*, vol. 19, no. 2, http://www.watershedsentinel.ca/content/saving-caribou-bc
6. The David Suzuki Foundation, 2010, 'Protecting species that need it', http://www.davidsuzuki.org/issues/wildlife-habitat/science/endangered-species-legislation/left-off-the-list-1/
7. Environment Canada, viewed 7 February 2013, Species at Risk Act, http://www.ec.gc.ca/alef-ewe/default.asp?lang=en&n=ED2FFC37-1
8. Nathan Vanderklippe, 16 May 2012, 'Reviving Arctic oil rush, Ottawa to auction rights in massive area', *The Globe and Mail*, http://www.theglobeandmail.com/news/politics/reviving-arctic-oil-rush-ottawa-to-auction-rights-in-massive-area/article4184419/

I. RAUCOUS SUMMER

1. J. G. Ballard, 2006, *Kingdom Come*, Fourth Estate, London.
2. *Hamlet*, Act 3, Scene 1.
3. T. S. Eliot, 1922, 'The Waste Land', Part 5.
4. *Chambers*, 12th edition.
5. Oliver Rackham, no date given, 'Ancient forestry practices', in Victor R. Squires (ed.), *The Role of Food, Agriculture, Forestry and Fisheries in Human Nutrition*, vol. II, *Encyclopedia of Life Support Systems*.
6. Dick Mol, John de Vos and Johannes van der Plicht, 2007, 'The presence and extinction of *Elephas antiquus* Falconer and Cautley, 1847, in Europe', *Quaternary International*, vols. 169–70, pp. 149–53.
7. 'Will-of-the-Land: Wilderness among Primal Indo-Europeans', *Environmental Review*, Winter 1985, vol. 9, no. 4, pp. 323–9.
8. George Byron, 1818, 'Childe Harold's Pilgrimage', Verse 178.
9. Christopher Smith, 1992, 'The population of Mesolithic Britain', *Mesolithic Miscellany*, vol. 13, no. 1.
10. Ibid. Smith estimates that Britain, towards the end of the Mesolithic, covered 270,000 km². The land area diminished as sea levels rose (it now stands at 230,000 km²).

2. THE WILD HUNT

1. Severin Carrell, 24 February 2012, 'Fishing skippers and factory fined nearly £1m for illegal catches', *Guardian*, http://www.guardian.co.uk/environment/2012/feb/24/fishing-skippers-fined-illegal-catches
2. See George Monbiot, 8 August 2011, 'Mutually assured depletion', http://www.monbiot.com/2011/08/08mutually-assured-depletion/
3. Lewis Smith, 1 March 2011, 'Spanish mackerel fleet penalised for quota-busting',http://www.fish2fork.com/news-index/Spanish-mackerel-fleet-penalised-for-quota-busting.aspx
4. See Winston Evans, quoted on the Newquay site, The Seafood of Cardigan Bay, http://www.newquay-westwales.co.uk/seafood.htm
5. European Environment Agency, 2011, 'State of commercial fish stocks in North East Atlantic and Baltic Sea', http://www.eea.europa.eu/data-and-maps/figures/state-of-commercial-fish-stocks-in-n-e-atlantic-and-baltic-sea-in

3. FORESHADOWINGS

1. Christopher Mitchelmore, 2010, 'Newfoundland & Labrador cod fishery', http://liveruralnl.com/2010/07/17/newfoundland-labrador-cod-fishery/
2. Martin Bell, 2007, *Prehistoric Coastal Communities: The Mesolithic in Western Britain. CBA Research Report 149*, Council for British Archaeology.
3. Ibid.
4. Royal Society for the Protection of Birds, 2009, *The Great Crane Project*, http://www.rspb.org.uk/supporting/campaigns/greatcraneproject/project. aspx
5. BBC, 3 September 2009, 'Cranes to breed on the levels', http://news.bbc. co.uk/local/somerset/hi/people_and_places/nature/newsid_8235000/8235479.stm
6. Bell, *Prehistoric Coastal Communities*.
7. Ibid.

4. ELOPEMENT

1. Benjamin Franklin, 9 May 1753, 'The support of the poor', letter to Peter Collinson, http://www.historycarper.com/resources/twobf2/letter18.htm
2. George Percy, quoted by David E. Stannard, 1992, *American Holocaust: The Conquest of the New World*, Oxford University Press, New York.
3. J. Hector St John de Crèvecoeur, 1785, *Letters from an American Farmer and Other Essays. Letter 12*, edited by Dennis D. Moore, Harvard University Press, Cambridge, MA.

5. THE NEVER-SPOTTED LEOPARD

1. Sion Morgan, 12 December 2010, 'Pembrokeshire "panther" strikes again', *Wales On Sunday*, http://www.walesonline.co.uk/news/wales-news/2010/12/12/pembrokeshire-panther-strikes-again-91466-27810028/
2. Ibid.
3. Mark Lingard, 29 January 2011, 'Big cat sighting in west Wales "100% authentic"', *County Times*.

4. Translation by Robert Williams, http://www.mythiccrossroads.com/PaGur.htm

5. Merrily Harpur, 2006, *Mystery Big Cats*, Heart of Albion, Market Harborough.

6. Mark Kinver, 30 October 2008, 'Snow leopard wins top photo prize', http://news.bbc.co.uk/l/hi/sci/tech/7696188.stm

7. Harpur, *Mystery Big Cats*.

8. S. J. Baker and C. J. Wilson, 1995, *The Evidence for the Presence of Large Exotic Cats in the Bodmin Area and their Possible Impact on Livestock*, a report by ADAS on behalf of the Ministry of Agriculture Fisheries and Food, http://www.naturalengland.org.uk/Images/exoticcats_tcm6-4645.pdf

9. No named author, 25 January 2010, 'Is the big cat mystery finally solved? Villagers find huge paw prints in snow after 30 years of sightings', *Daily Mail*, http://www.dailymail.co.uk/news/article-1245816/Is-big-cat-mystery-solved-Villagers-huge-paw-prints-snow-30-years-sightings.html

10. No named author, 10 January 2011, 'Do giant paw prints mean big cat is on the prowl in capital?', *Scotsman*, http://www.scotsman.com/news/do-giant-paw-prints-mean-big-cat-is-on-the-prowl-in-capital-1-1489992

11. Patrick Barkham, 23 March 2005, 'Fear stalks the streets of Sydenham after resident is attacked by a black cat the size of a labrador', *Guardian*, http://www.guardian.co.uk/uk/2005/mar/23/patrickbarkham

12. BBC, 22 March 2005, '"Big cat" attacks man in garden', http://news.bbc.co.uk/1/hi/england/london/4370893.stm

13. Paul Harris, 9 January 2009, 'Is this the Beast of Exmoor? Body of mystery animal washes up on beach', *Daily Mail*, http://www.dailymail.co.uk/news/article-1109174/Is-Beast-Exmoor-Body-mystery-animal-washes-beach.html

14. David Hambling, 2001, 'How big is an alien big cat?', *The Skeptic*, vol. 14, no. 4, pp. 8–11.

15. Richard Wiseman, 2011, *Paranormality: Why We See What Isn't There*, Macmillan, London.

16. See, for example: Dominic Sandbrook, 17 July 2010, 'A perfect folk hero for our times', *Daily Mail*, http://www.dailymail.co.uk/debate/article-1295459/A-perfect-folk-hero-times-Moat-popularity-reflects-societys-warped-values.html; Emily Andrews, Daniel Martin and Paul Sims, 16 July 2010, 'I set up the Moat Facebook tributes: the single mother behind twisted online shrine', *Daily Mail*, http://www.dailymail.co.uk/news/article-1295141/Siobhan-ODowd-set-Raoul-Moat-Face book-tribute-site.html; John Demetriou, 16 July 2010, 'Raoul Moat: sympathy

for the devil?', http://www.boatangdemetriou.com/2010/07/raoul-moat-sympathy-for-devil.html

6. GREENING THE DESERT

1. www.cambrian-mountains.co.uk/
2. Graham Uney, 1999, *The High Summits of Wales*, Logaston Press, Hereford. Quoted by the Cambrian Mountains Society, http://www. cambrian-mountains.co.uk/documents/cambrian-mountains-sustainable-future-low-graphics.pdf
3. Fiona R. Grant, 2009, *Analysis of a Peat Core from the Clwydian Hills, North Wales*. Report produced for Royal Commission on the Ancient and Historical Monuments of Wales, http://www.rcahmw.gov.uk/media/193.pdf
4. Ibid.
5. See for example, R. Fyfe, 2007, 'The importance of local-scale openness within regions dominated by closed woodland', *Journal of Quaternary Science*, vol. 22, no. 6, pp. 571–8, doi: 10.1002/jqs.1078; J. H. B. Birks, 2005, 'Mind the gap: how open were European primeval forests?', *Trends in Ecology & Evolution*, vol. 20, pp. 154–6.
6. Richard Tyler, 17 December 2007, quoted in the *Western Mail*.
7. Countryside Council for Wales, 2011, 'Claerwen', http://www.ccw.gov. uk/landscape--wildlife/protecting-our-landscape/special-landscapes--sites/protected-landscapes/national-nature-reserves/claerwen.aspx
8. Daniel Pauly, 1995, 'Anecdotes and the shifting baseline syndrome of fisheries', *Trends in Ecology & Evolution*, vol. 10, no. 10, doi: 10.1016/S0169-5347(00)89171-5.
9. Derek Yalden, 1999, *The History of British Mammals*, T and AD Poyser, London.
10. R. C. Tassell, 2011, *Direct Sowing of Birch on an Upland Dense Bracken Site, 2002–2011*, Coed Cymru, Powys.
11. Trees for Life, no date given, 'Seed dispersal', http://www.treesforlife. uk/forest/ecological/seed_dispersal.html
12. Bryony Coles, 2006, *Beavers in Britain's Past*, Oxbow Books and WARP, Oxford.
13. Ibid.
14. Derek Gow, 2006, 'Beaver trends in Britain and Europe', *ECOS*, vol. 27, no. 1, pp. 57–65.
15. Ibid.
16. Ibid.

17. Severin Carrell, 25 November 2010, 'Scotland's beaver-trapping plan has wildlife campaigners up in arms', *Guardian*, http://www.guardian.co.uk/environment/2010/nov/25/beavers-scotland-conservation

18. Richard Vaughan, FUW, quoted by Sally Williams, 8 April 2011, 'Beavers scheme just "crazy", farmers warn', *Western Mail*.

19. The Blaeneinion Project, 2011, *Beaver Fact Sheet*, http://www.blaeneinion.co.uk

20. William J. Ripple and Robert L. Beschta, 2012, 'Trophic cascades in Yellowstone: the first 15 years after wolf reintroduction', *Biological Conservation*, vol. 145, issue 1, pp. 205–13.

21. Åsa Hägglund and Göran Sjöberg, 1999, 'Effects of beaver dams on the fish fauna of forest streams', *Forest Ecology and Management*, vol. 115, nos. 2–3, pp. 259–66, doi:10.1016/S0378-1127(98)00404-6; Krzysztof Kukula and Aneta Bylak, 2010, 'Ichthyofauna of a mountain stream dammed by beaver', *Archives of Polish Fisheries*, vol. 18, no. 1, pp. 33–43, doi: 10.2478/v10086-010-0004-1.

22. Douglas B. Sigourney et al, 2006, 'Influence of beaver activity on summer growth and condition of age-2 Atlantic salmon parr', *Transactions of the American Fisheries Society*, vol. 135, no. 4, pp. 1068–75, doi: 10.1577/T05-159.1.

23. Robert J. Naiman, Carol A. Johnston and James C. Kelley, 1988, 'Alteration of North American streams by beaver', *BioScience*, vol. 38, no. 11, pp. 753–62, http://www.jstor.org/stable/1310784

24. Mateusz Ciechanowski et al, 2011, 'Reintroduction of beavers *Castor fiber* may improve habitat quality for vespertilionid bats foraging in small river valleys', *European Journal of Wildlife Research*, vol. 57, pp. 737–47, doi: 10.1007/s10344-010-0481-y.

25. Nick Mott, 2005, *Managing Woody Debris in Rivers and Streams*, Water for Wildlife and the Wildlife Trusts, http://www.riou.be/pdf/extern/Woody%20Debris%20Booklet.pdf

26. See fig. 20.22, chap. 20 of the 'UK national ecosystem assessment', http://uknea.unep-wcmc.org/Resources/tabid/82/Default.aspx

27. Mott, *Managing Woody Debris*.

28. Forest Research, 2012, *Slowing the Flow at Pickering: What is the Project?*, http://www.forestry.gov.uk/fr/INFD-7ZUCL6; Forest Research, *Slowing the Flow in Pickering and Sinnington*, http://www.forestry.gov.uk/pdf/Slow_the_flow_Pickering_factsheet.pdf/$FILE/Slow_the_flow_Pickering_factsheet.pdf

29. Naiman, Johnston and Kelley, 'Alteration of North American streams by beaver'.

30. Sally Williams, 8 April 2011, 'Beavers scheme just "crazy", farmers warn', *Western Mail.*
31. Quentin D. Skinner et al, 1984, 'Stream water quality as influenced by beaver within grazing systems in Wyoming', *Journal of Range Management*, vol. 37, no. 2, pp. 142–6.
32. Ripple and Beschta, 'Trophic cascades in Yellowstone'.
33. Ibid.
34. R. J. Naiman and K. H. Rogers, 1997, 'Large animals and system-level characteristics in river corridors', *BioScience*, 47, p. 521, doi: 10.2307/1313120. Cited in J. A. Estes et al, 2011, 'Trophic downgrading of planet earth', *Science*, vol. 333, no. 6040, pp. 301–6, doi: 10.1126/science.1205106.
35. Lisa Marie Baril, 2009, 'Change in deciduous woody vegetation, implications of increased willow (*Salix* Spp.) growth for bird species diversity, and willow species composition in and around Yellowstone National Park's Northern Range', MSc thesis, Montana State University, http://etd.lib.montana.edu/etd/2009/baril/BarilL1209.pdf
36. Ripple and Beschta, 'Trophic cascades in Yellowstone'.
37. Robert L. Beschta and William J. Ripple, 2006, 'River channel dynamics following extirpation of wolves in northwestern Yellowstone National Park', *Earth Surface Processes and Landforms*, vol. 31, no. 12, pp. 1525–39, doi: 10.1002/esp.1362.
38. William J. Ripple and Robert L. Beschta, 2006, 'Linking a cougar decline, trophic cascade, and catastrophic regime shift in Zion National Park', *Biological Conservation*, vol. 133, pp. 397–408, doi: 10.1016/j.biocon.2006.07.002.
39. Robert L. Beschta and William J. Ripple, 2009, 'Large predators and trophic cascades in terrestrial ecosystems of the western United States', *Biological Conservation*, vol. 142, pp. 2401–14.
40. Douglas A. Frank, 2008, 'Evidence for top predator control of a grazing ecosystem', *Oikos*, 117, pp. 1718–24, doi: 10.1111/j.1600-0706.2008.16846.x.
41. Ripple and Beschta, 'Trophic cascades in Yellowstone'.
42. Ibid.
43. Beschta and Ripple, 'Large predators and trophic cascades'.
44. Adrian D. Manning, Iain J. Gordon and William J. Ripple, 2009, 'Restoring landscapes of fear with wolves in the Scottish Highlands', *Biological Conservation*, vol. 142, issue 10, pp. 2314–21, http://dx.doi.org/10.1016/j.biocon.2009.05.007

45. G. V. Hilderbrand, et al, 1999, 'Role of brown bears (*Ursus arctos*) in the flow of marine nitrogen into a terrestrial ecosystem', *Oecologia*, 121, pp. 546–50.

46. D. A. Croll et al, 2005, 'Introduced predators transform subarctic islands from grassland to tundra', *Science*, vol. 30, 7, no. 5717, pp. 1959–61, doi: 10.1126/science.1108485.

47. S. A. Zimov et al, 1995, 'Steppe-tundra transition: a herbivore-driven biome shift at the end of the Pleistocene', *The American Naturalist*, vol. 146, no. 5, pp. 765–94.

48. See D. Nogués-Bravo et al, 2008, 'Climate change, humans, and the extinction of the woolly mammoth', *PLoS Biology*, vol. 6, no. 4, p. 79, doi: 10.1371/journal.pbio.0060079.

49. Zimov et al, 'Steppe-tundra transition'.

50. S. A. Zimov, 2005, 'Pleistocene Park: return of the mammoth's ecosystem', *Science*, vol. 308, pp. 796–8, doi: 10.1126/science.1113442.

51. Zimov et al, 'Steppe-tundra transition'.

52. Susan Rule et al, 2012, 'The aftermath of megafaunal extinction: ecosystem transformation in Pleistocene Australia', *Science*, vol. 335, pp. 1483–6, doi: 10.1126/science.1214261.

53. Laura R. Prugh et al, 2009, 'The rise of the Mesopredator', *BioScience*, 59(9), pp. 779–91.

54. James A. Estes et al, 2011, 'Trophic downgrading of planet earth', *Science*, vol. 333, no. 6040, pp. 301–6, doi: 10.1126/science.1205106.

55. Prugh et al, 'The rise of the Mesopredator'.

56. Anil Markandya et al, 2008, 'Counting the cost of vulture decline – an appraisal of the human health and other benefits of vultures in India', *Ecological Economics*, 67 (2), pp. 194–204.

7. BRING BACK THE WOLF

1. Dick Mol, John de Vos and Johannes van der Plicht, 2007, 'The presence and extinction of *Elephas antiquus* Falconer and Cautley, 1847, in Europe', *Quaternary International*, vols. 169–70, pp. 149–53.

2. S. L. Vartanyan, V. E. Garutt and A. V. Sher, 1993, 'Holocene dwarf mammoths from Wrangel Island in the Siberian Arctic', *Nature*, vol. 362, pp. 337–40, doi: 10.1038/362337a0.

3. A. J. Stuart, 2001, 'Occurrence of mammalia relicts at site Trafalgar Square', *European Quaternary Mammalia Database*, http://doi.pangaea.de/10.1594/PANGAEA.64391; J. W. Franks, 1959, 'Interglacial depos-

its at Trafalgar Square, London', re-issued 2006, in *New Phytologist*, vol. 59, issue 2.

4. Hervé Bocherens et al, 2011, 'Isotopic evidence for dietary ecology of cave lion (*Panthera spelaea*) in north-western Europe: prey choice, competition and implications for extinction', *Quaternary International*, vol. 245, no. 2, pp. 249–61, http://dx.doi.org/10.1016/j.quaint.2011.02.023

5. Derek Yalden, 1999, *The History of British Mammals*, T and AD Poyser, London.

6. Mary C. Stiner, 2004, 'Comparative ecology and taphonomy of spotted hyenas, humans, and wolves in Pleistocene Italy', *Revue de Paléobiologie*, vol. 23, no. 2, pp. 771–85.

7. Franks, 'Interglacial deposits at Trafalgar Square'.

8. Oliver Rackhan, no date given, 'Ancient forestry practices', in Victor R. Squires (ed.), *The Role of Food, Agriculture, Forestry and Fisheries in Human Nutrition*, vol. II, *Encyclopedia of Life Support Systems*.

9. Jonas Chafota, 1998, 'Effects of changes in elephant densities on the environment and other species: how much do we know?', Cooperative Regional Wildlife Management in Southern Africa, http://agecon.ucdavis. edu/people/faculty/lovell-jarvis/docs/elephant/chafota.pdf; J. J. Smallie and T. G. O'Connor, 2000, 'Elephant utilization of *Colophospermum mopane*: possible benefits of hedging', *African Journal of Ecology*, vol. 38, pp. 352–9.

10. Graham Kerley et al, 2008, 'Effects of elephants on ecosystems and biodiversity', in R. J. Scholes and K. G. Mennell (eds.), *Elephant Management: A Scientific Assessment of South Africa*, Witwatersrand University Press, Johannesburg; Peter Baxter, 2003, 'Modeling the impact of the African elephant, *Loxodonta africana*, on woody vegetation in semi-arid savannas', PhD dissertation, University of California, Berkeley.

11. Department for Environment, Food and Rural Affairs, 2008, 'Feral wild boar in England: an action plan', http://www.naturalengland.org.uk/ Images/feralwildboar_tcm6-4508.pdf

12. M. J. Goulding and T. J. Roper, 2002, 'Press responses to the presence of free-living wild boar (*Sus scrofa*) in southern England', *Mammal Review*, 32, pp. 272–82, doi: 10.1046/j.1365-2907.2002.00109.x.

13. Department for Environment, Food and Rural Affairs, 'Feral wild boar in England'.

14. Derek Gow, 2002, 'A wallowing good time – wild boar in the woods', *ECOS*, 23 (2), pp. 14–22.

15. Trees for Life, 2008, 'Results from the Guisachan Wild Boar Project', http://www.treesforlife.org.uk/forest/missing/guisachan200805.html

16. Department for Environment, Food and Rural Affairs, 'Feral wild boar in England'.

17. Camila Ruz, 1 September 2011, 'Wild boar cull "not based on scientific estimates"', http://www.guardian.co.uk/environment/2011/sep/01/wild-boar-cull

18. Gow, 'A wallowing good time'.

19. British Wild Boar Organisation, January 2010, 'Interesting happenings occurring with Britain's free-living wild boar', http://www.britishwild-boar.org.uk/BWBONewsletterJan2010.pdf

20. Jenny Farrant, 2 February 2012, by email.

21. Northern Potential, 2011, 'The Highlands of Scotland', http://northern-potential.net/the_highlands_of_scotland

22. Alan Watson Featherstone, 2001, 'The wild heart of the Highlands', Trees for Life, http://www.treesforlife.org.uk/tfl.wildheart.html

23. Land Reform (Scotland) Act 2003, http://www.legislation.gov.uk/asp/2003/2/contents

24. Peter Fraser, Angus MacKenzie and Donald MacKenzie, 2012, 'The economic importance of red deer to Scotland's rural economy and the political threat now facing the country's iconic species', Scottish Game-keepers' Association.

25. Ibid.

26. BBC Scotland, 16 June 2011, 'Mull's economy soars on wings of white-tailed eagles', http://www.bbc.co.uk/news/uk-scotland-scotland-business-13783555

27. RSPB, various dates, 'Mull white-tailed eagles', http://www.rspb.org.uk/wildlife/tracking/mulleagles/

28. BBC Scotland, 'Mull's economy soars on wings of white-tailed eagles'.

29. The Scottish Government, June 2010, *The Economic Impact of Wildlife Tourism in Scotland*, research conducted by International Centre for Tourism and Hospitality Research, Bournemouth University, http://www.scotland.gov.uk/Publications/2010/05/12164456/1

30. Patrick Barkham, 14 September 2011, 'Record numbers of golden eagles poisoned in Scotland in 2010', http://www.guardian.co.uk/environment/2011/sep/14/golden-eagles-poisoned-scotland-rspb

31. Alan Watson Featherstone, 2010, 'Restoring biodiversity in the native pinewoods of the Caledonian Forest', *Reforesting Scotland*, issue 41, pp. 17–21, http://www.treesforlife.org.uk/images/Reforesting%20Scotland%2041%20Biodiversity.pdf

32. Dan Puplett, no date given, 'Riparian woodlands', http://www.treesforlife.org.uk/forest/ecological/riparianwoodland.html

33. David Hetherington, 13 July 2010, presentation at Re wilding Europe and the Return of Predators. Symposium convened by the Zoological Society of London.

34. Kevin Cahill, 2002, *Who Owns Britain*, Canongate, Edinburgh.

35. Rewilding Europe, 2012, *Making Europe a Wilder Place*, www. rewildingeurope.com/assets/uploads/Downloads/Rewilding-Europe-Brochure-2012.pdf

36. Rewilding Europe, 2012, 'First wild bison in Romania after 160 years', http://rewildingeurope.com/news/articles/first-wild-bison-in-romania-after-160-years/

37. http://www.panparks.org/newsroom/news/2012/wilderness-does-not-stop-at-borders

38. WWF, 2012, 'Danube-Carpathian region', http://wwf.panda.org/what_we_do/where_we_work/black_sea_basin/danube_carpathian/blue_river_green_mtn/; Wild Europe, 2010, 'Towards a wilder Europe: developing an action agenda for wilderness and large natural habitat areas', http://www.panparks.org/sites/default/files/docs/publications-resources/towards_a_wilder_europe.pdf

39. Wild Europe, 2010, Restoration Conference, http://www.wildeurope.org/index.php?option=com_content&view=article&id=56&Itemid=19

40. Wild Europe, 'Towards a wilder Europe'.

41. Pan Parks, 27 June 2012, Genuine wilderness protection in Germany', http://www.panparks.org/newsroom/news/2012/genuine-wilderness-protection-in-germany

42. Twan Teunissen, 3 October 2011, 'Horses to the wolves, wolves to the horses', http://rewildingeurope.com/blog/horses-to-the-wolves-wolves-to-the-horses/

43. Ibid.

44. Suzanne Goldenberg, 8 December 2010, 'How America is learning to live with wolves again', http://www.guardian.co.uk/environment/2010/dec/08/keeping-wolf-from-door

45. Wildlife Extra, September 2011, 'Wolf caught on camera trap in Belgium' (Video), www.wildlifeextra.com/go/news/wolf-belgium.html

46. Erwin van Maanen, 2011, 'Wolves marching further west!', http://www.rewildingfoundation.org/2011/09/23/wolves-marching-further-west/

47. Rewilding Europe, *Making Europe a Wilder Place*.

48. Ibid.

49. International Union for Conservation of Nature, 2013. IUCN Red List of Threatened Species: *Bison bonasus*. http://www.iucnredlist.org/details/2814/0

50. The Blaeneinion Project, 2011, *Beaver Fact Sheet*, http://www. blaeneinion.co.uk

51. Rewilding Europe, *Making Europe a Wilder Place*.

52. J. D. C. Linnell et al, 2002, 'The fear of wolves: a review of wolf attacks on humans', NINA (Norwegian Institute for Nature Research) Opp dragsmelding 731, http://www.nina.no/archive/nina/PppBasePdf/ oppdragsmelding/731.pdf

53. Roger Panaman, 2002, 'Wolves are returning', *ECOS*, vol. 23, no. 2.

54. US Fish and Wildlife Service, 1993, 'The reintroduction of gray wolves to Yellowstone National Park and Central Idaho: environmental impact statement', Gray Wolf Environmental Impact Study, Helena, Montana. Cited by Panaman, 'Wolves are returning'.

55. P. Ciucci and L. Boitani, 1998, 'Wolf and dog depredation on livestock in central Italy', *Wildlife Society Bulletin*, vol. 26, pp. 504–14.

56. Laetitia M. Navarro and Henrique M. Pereira, 2012, 'Rewilding abandoned landscapes in Europe', *Ecosystems*, vol. 15, no. 6, pp. 900–912.

57. Charles J. Wilson, 2004, 'Could we live with reintroduced large carnivores in the UK?', *Mammal Review*, vol. 34, no. 3, pp. 211–32.

58. BBC Technology, 6 August 2012, 'Sheep to warn of wolves via text message', www.bbc.co.uk/news/technology-19147403

59. Guillaume Chapron, 13 July 2010, 'Restoring and managing wolves in Sweden', presentation at Rewilding Europe and the Return of Predators. Symposium convened by the Zoological Society of London.

60. Oliver Rackham, 1986, *The History of the Countryside*, JM Dent and Sons, London.

61. Wilson, 'Could we live with reintroduced large carnivores in the UK?'; Panaman, 'Wolves are returning'.

62. Erlend B. Nilsen et al, 2007, 'Wolf reintroduction to Scotland: public attitudes and consequences for red deer management', *Proceedings of the Royal Society – B*, vol. 274, no. 1612, pp. 995–1003, doi: 10.1098/ rspb.2006.0369.

63. Ibid.

64. D. P. J. Kuijper, 2011, 'Lack of natural control mechanisms increases wildlife–forestry conflict in managed temperate European forest systems', *European Journal of Forest Research*, vol. 130, no. 6, pp. 895–909, doi: 10.1007/s10342-011-0523-3.

65. Dan Puplett, 2008, 'Our once and future fauna', *ECOS*, vol. 29, pp. 4–17.

66. Laura R. Prugh et al, 2009, 'The rise of the Mesopredator', *BioScience*, vol. 59, no. 9, pp. 779–91.

67. Nilsen et al, 'Wolf reintroduction to Scotland'.

68. R. D. S. Jenkinson, 1983, 'The recent history of Northern Lynx (*Lynx lynx* Linne) in the British Isles', *Quaternary Newsletter*, vol. 41, pp. 1–7. Cited in David A. Hetherington, Tom C. Lord and Roger M. Jacobi, 2006, 'New evidence for the occurrence of Eurasian lynx (*Lynx lynx*) in medieval Britain', *Journal of Quaternary Science*, vol. 21, no. 1, pp. 3–8, doi: 10.1002/jqs.960.

69. Hetherington, Lord and Jacobi, 'New evidence for the occurrence of Eurasian lynx (*Lynx lynx*) in medieval Britain'.

70. http://www.cs.ox.ac.uk/people/geraint.jones/rhydychen.org/about.welsh/pais-dinogad.html

71. Darren Devine, 12 October 2005, 'Was Welsh poet right about lynx legend?', *Western Mail*, www.walesonline.co.uk/news/wales-news/tm_objectid=16238211&method=full&siteid=50082&headline=was-welsh-poet-right-about-lynx-legend–name_page.html

72. David Hetherington, 2010, 'The lynx', in Terry O'Connor and Naomi Sykes (eds.), *Extinctions and Invasions: A Social History of British Fauna*, Windgather Press, Oxford.

73. Wilson, 'Could we live with reintroduced large carnivores in the UK?'.

74. David Hetherington, 13 July 2010, 'The potential for restoring Eurasian lynx to Scotland', presentation at Rewilding Europe and the Return of Predators. Symposium convened by the Zoological Society of London.

75. U. Breitenmoser et al, 2000, *The Action Plan for the Conservation of the Eurasian Lynx (Lynx Lynx) in Europe*, Council of Europe Publishing, Strasbourg, France, *Nature and Environmental Series No. 112*. Cited by David Hetherington et al, 2008, 'A potential habitat network for the Eurasian lynx *Lynx lynx* in Scotland', *Mammal Review*, vol. 38, no. 4, pp. 285–303.

76. David Hetherington, 2006, 'The lynx in Britain's past, present and future', *ECOS*, vol. 27, no. 1, pp. 66–74.

77. Hetherington et al, 'A potential habitat network for the Eurasian lynx *Lynx lynx* in Scotland'.

78. Hetherington, 'The potential for restoring Eurasian lynx to Scotland'.

8. A WORK OF HOPE

1. Bryony Coles, 2006, *Beavers in Britain's Past*, Oxbow Books and WARP, Oxford.

2. Oliver Rackham, 1986, *The History of the Countryside*, JM Dent and Sons, London.

3. Derek Yalden, 1999, *The History of British Mammals*, T and AD Poyser, London.

4. Ibid.

5. The Cairngorm Reindeer Herd, various dates, www.cairngormreindeer. co.uk/

6. R. Coard and A. T. Chamberlain, 1999, 'The nature and timing of faunal change in the British Isles across the Pleistocene/Holocene transition', *The Holocene*, 9, p. 372, doi: 10.1191/095968399672435429; Yalden, *British Mammals*.

7. BIAZA, 2012, 'Eelmoor Marsh Conservation Project', www.biaza.org.uk/ conservation/conservation-projects/eelmoor-marsh-conservation-project/

8. Yalden, *British Mammals*.

9. David Hetherington, 2010, 'The lynx', in Terry O'Connor and Naomi Sykes (eds.), *Extinctions and Invasions: A Social History of British Fauna*, Windgather Press, Oxford.

10. Rackham, *History of the Countryside*.

11. Ibid.

12. The Mammal Society, 2011, www.mammal.org.uk/index.php?option=com_ content&view=article&id=250&Itemid=283

13. Ibid.

14. Yalden, *British Mammals*.

15. Mary C. Stiner, 2004, 'Comparative ecology and taphonomy of spotted hyenas, humans, and wolves in Pleistocene Italy', *Revue de Paléobiologie*, vol. 23, no. 2, pp. 771–85.

16. Dick Mol, John de Vos and Johannes van der Plicht, 2007, 'The presence and extinction of *Elephas antiquus* Falconer and Cautley, 1847, in Europe', *Quaternary International*, vols. 169–70, pp. 149–53.

17. Yalden, *British Mammals*.

18. Ibid.

19. No author given, 18 July 2005, 'Plan to bring grey whales back to Britain', *Daily Telegraph*, www.telegraph.co.uk/news/uknews/1494286/ Plan-to-bring-grey-whales-back-to-Britain.html

20. Yalden, *British Mammals*.

21. J. R. Waldman, 2000, 'Restoring *Acipenser sturio* L., 1758 in Europe: lessons from the *Acipenser oxyrinchus* Mitchill, 1815 experience in North America', *Boletín, Instituto Español de Oceanografía*, vol. 16, pp. 237–44.

22. Jörn Gessner et al, 2006, 'Remediation measures for the Baltic sturgeon:

status review and perspectives', *Journal of Applied Ichthyology*, vol. 22, issue supplement s1, pp. 23–31, doi: 10.1111/j.1439-0426.2007.00925.x; F. Kirschbaum and J. Gessner, 2000, 'Re-establishment programme for *Acipenser sturio* L. 1758: the German approach', *Boletín, Instituto Español de Oceanografía*, vol. 16, pp. 149–56.

23. P. Williot et al, 2009, '*Acipenser sturio* recovery research actions in France', in *Biology, Conservation and Sustainable Development of Sturgeons, Fish & Fisheries Series*, vol. 29, III, pp. 247–63, Springer, Germany, doi: 10.1007/978-1-4020-8437-9_15.

24. Mull Magic, 2012. 'White-tailed eagles on the Isle of Mull', www.white-tailed-sea-eagle.co.uk/

25. British Birds, 1 August 2010, 'White-tailed eagle reintroduction grounded', www.britishbirds.co.uk/news-and-comment/white-tailed-eagle-reintroduction-grounded

26. Dyfi Osprey Project, 2011, 'History of British ospreys', www.dyfiospreyproject.com/history-of-british-ospreys

27. Tim Melling, Steve Dudley and Paul Doherty, 2008, 'The eagle owl in Britain', *British Birds*, vol. 101, pp. 478–90.

28. D. W. Yalden and U. Albarella, 2009, *The History of British Birds*, Oxford University Press, Oxford.

29. Ibid.

30. Royal Society for the Protection of Birds, 2012, *Goshawk*, www.rspb.org.uk/wildlife/birdguide/name/g/goshawk/index.aspx

31. Forestry Commission, 2012, *Capercaillie*, www.forestry.gov.uk/forestry/capercaillie

32. Trees for Life, 1999, *Species Profile: Capercaillie*, www.treesforlife.org.uk/tfl.capercaillie.html

33. Clive Hambler and Susan M. Canney, 2013 (2nd edition, read in galley proof), *Conservation*, Cambridge University Press, Cambridge.

34. Wildlife Extra, 2007, 'Great Bustards in the UK', www.wildlifeextra.com/go/news/bw-greatbustards.html

35. Andrew Stanbury and the UK Crane Working Group, 1 August 2011, 'The changing status of the common crane in the UK', www.britishbirds.co.uk/articles/the-changing-status-of-the-common-crane-in-the-uk

36. Peter Taylor, 2011, 'Big birds in the UK: the reintroduction of iconic species', *ECOS*, vol. 32, no. 1, pp. 74–80.

37. Yalden and Albarella, *British Birds*.

38. BBC News, 23 April 2004, 'Storks set to end 600-year wait', http://news.bbc.co.uk/1/hi/england/west_yorkshire/3653171.stm

39. Royal Society for the Protection of Birds, 2012, 'Something to stork about!', www.rspb.org.uk/community/wildlife/b/wildlife/archive/2012/04/26/something-to-stork-about.aspx

40. Yalden and Albarella, *British Birds*.

41. Natural England, 12 September 2011, 'Breeding spoonbills return to Holkham', www.naturalengland.org.uk/about_us/news/2011/120911.aspx

42. Natural England, 21 November 2012, by email.

43. Yalden and Albarella, *British Birds*.

44. Ibid.

45. Ibid.

46. See S. A. Zimov et al, 1995, 'Steppe–tundra transition: a herbivore-driven biome shift at the end of the Pleistocene', *The American Naturalist*, vol. 146, no. 5, pp. 765–94.

47. See S. A. Zimov, 2005, 'Pleistocene Park: return of the mammoth's ecosystem', *Science*, vol. 308, pp. 796–8, doi: 10.1126/science. 1113442.

48. www.riverbluffcave.com/gallery/rec_id/104/type/1

49. Nancy Sisinyak, no date given, 'The biggest bear ... ever', Alaska Department of Fish and Game, www.adfg.alaska.gov/index.cfm?adfg= wildlifenews.view_article&articles_id=232&issue_id=41

50. San Diego Zoo, April 2009, *Extinct Teratorn, Teratornithidae*, http://library.sandiegozoo.org/factsheets/_extinct/teratorn/teratorn.htm

51. For example, Paul S. Martin, 2005, *Twilight of the Mammoths: Ice Age Extinctions and the Rewilding of America*, University of California Press, Berkeley; F. L. Koch and A. D. Barnosky, 2006, 'Late Quaternary extinctions: state of the debate', *Annual Review of Ecology, Evolution, and Systematics*, vol. 37, pp. 215–50.

52. See William J. Ripple and Blaire Van Valkenburgh, 2010, 'Linking top-down forces to the Pleistocene megafaunal extinctions', *BioScience*, vol. 60, no. 7, pp. 516–26, doi: 10.1525/bio.2010.60.7.7.

53. See Ripple and Van Valkenburgh, 'Linking top-down forces'.

54. Josh Donlan et al, 2005, 'Re-wilding North America', *Nature*, vol. 436, pp. 913–14, doi: 10.1038/436913a; Tim Caro, 2007, 'The Pleistocene re-wilding gambit', *Trends in Ecology & Evolution*, vol. 22, no. 6, pp. 281–3, doi: 10.1016/j.tree.2007.03.001.

55. Dustin R. Rubenstein et al, 2006, 'Pleistocene Park: does re-wilding North America represent sound conservation for the 21st century?', *Biological Conservation*, vol. 132, pp. 232–8, doi: 10.1016/j.biocon. 2006.04.003.

56. Peter Taylor, 2009, 'Re-wilding the grazers: obstacles to the "wild" in wildlife management', *British Wildlife*, vol. 51, no. 5 (special supplement), pp. 50–55.

57. Pleistocene Park, various dates, www.pleistocenepark.ru/en/

58. Zimov, 'Pleistocene Park'.

59. www.pleistocenepark.ru/en/background/

60. Zimov et al, 'Steppe-tundra transition'.

61. Mike D'Aguillo, 2008, 'Recreating a wooly mammoth', http://sites. google.com/site/mikesbiowebpage/mammoth-recreation-project; Nicholas Wade, 9 November 2008, 'Regenerating a mammoth for $10 million', *New York Times*, www.nytimes.com/2008/11/20/science/20mammoth. html?pagewanted=all&_r=0

62. Global Invasive Species Database, 2012, '*Clarias batrachus*', www.issg. org/database/species/ecology.asp?si=62&fr=1&sts=sss&lang=EN

63. Global Invasive Species Database, 2012, '*Rhinella marina* (= *Bufo marinus*)', www.issg.org/database/species/ecology.asp?si=113&fr=1&sts =sss&lang=EN

64. John Vidal, 20 May 2008, 'From stowaway to supersize predator: the mice eating rare seabirds alive', *Guardian*, www.guardian.co.uk/ environment/2008/may/20/wildlife.endangeredspecies

65. Offwell Woodland & Wildlife Trust, 2011, 'The value of different tree species for invertebrates and lichens'. Data extracted from C. E. J. Kennedy and T. R. E. Southwood, 1984, 'The number of species of insects associated with British trees: a re-analysis', *Journal of Animal Ecology*, vol. 53, pp. 455–78, www.countrysideinfo.co.uk/woodland_manage/tree_value.htm

66. Christopher D. Preston, David A. Pearman and Allan R. Hall, 2004, 'Archaeophytes in Britain', *Botanical Journal of the Linnean Society*, vol. 145, pp. 257–94.

67. Jagjit Singh et al, 1994, 'The search for wild dry rot fungus (*Serpula lacrymans*) in the Himalayas', *Journal of the Institute of Wood Science*, vol. 13, no. 3, pp. 411–12.

68. Preston, Pearman and Hall, 'Archaeophytes in Britain'.

69. Ibid.

70. Christine M. Cheffings and Lynne Farrell (eds.), 2005, 'Species Status No. 7', *The Vascular Plant Red Data List for Great Britain*, Joint Nature Conservation Committee, http://jncc.defra.gov.uk/pdf/pub05_ speciesstatusvpredlist3_web.pdf

71. Plantlife, 2011, *Pheasant's-eye*, www.plantlife.org.uk/wild_plants/plant_ species/pheasants-eye

72. Yalden, *British Mammals*.

73. Book V, 12, cited by Yalden, *British Mammals*.

74. Forestry Commission, 29 July 2008, 'Goshawks are stars of the show at Haldon!', www.forestry.gov.uk/newsreel.nsf/WebPressReleases/016336 9508A2CD7380257488005 22E10; Rob Coope, 2007, 'A preliminary investigation of the food and feeding behaviour of pine martens in productive forestry from an analysis of the contents of their scats collected in Inchnacardoch forest, Fort Augustus', *Scottish Forestry*, vol. 61, no. 3, pp. 3–15.

75. Yalden, *British Mammals*.

76. P. Salo et al, 2008, 'Risk induced by a native top predator reduces alien mink movements', *Journal of Animal Ecology*, vol. 77, no. 6, pp. 1092–8, doi: 10.1111/j.1365-2656.2008.01430.x.

77. Guy Hand, October 2000, 'Planting on barren ground', Trees for Life, www.treesforlife.org.uk/tfl.guyhand.html

78. Dan Puplett, no date given, 'Dead wood', Trees for Life, www.treesforlife.org.uk/forest/ecological/deadwood.html

79. Alan Watson Featherstone, 2001, 'The wild heart of the Highlands', Trees for Life, www.treesforlife.org.uk/tfl.wildheart.html

80. Ibid.

9. SHEEPWRECKED

1. Woodland Trust, 2012, *UK Woodland Facts*, www.woodlandtrust.org. uk/en/news-media/fact-file/Pages/uk-woodland-facts.aspx#.Tp7vU3LDD90

2. Thomas More, *Utopia*, chapter 22.

3. David Williams, 1952, 'Rhyfel y Sais Bach: an enclosure riot on Mynydd Bach', *Journal of the Cardiganshire Antiquarian Society*, vol. 2, nos. 1–4.

4. National Library of Wales, 2004, 'Life on the land: land ownership', http://digidol.llgc.org.uk/METS/XAM00001/ardd?locale=en

5. In evidence submitted to the House of Commons Environment, Food and Rural Affairs Committee, 16 February 2011, 'Farming in the uplands', Third Report of Session 2010–11, http://www.publications. parliament.uk/pa/cm201011/cmselect/cmenvfru/556/556.pdf

6. Statistics for Wales, 2011, *Agricultural Small Area Statistics for Wales, 2002 to 2010*. SB 75/2011, http://wales.gov.uk/docs/statistics/2011/110 728sb752011en.pdf

7. UK National Ecosystem Assessment (2011), chap. 20, fig. 20.8, 'Short-term abundance of widespread breeding birds in Wales 1994–2009', http://uknea.unep-wcmc.org/Resources/tabid/82/Default.aspx

8. Royal Society for the Protection of Birds Cymru, 2009, Submission to Rural Development Sub-Committee Inquiry into the future of the uplands in Wales, http://www.assemblywales.org/6_rspb_formatted.pdf

9. UK National Ecosystem Assessment, chap. 20, fig. 20.16, 'Condition of a) riverine species, and b) riverine habitats in special areas of conservation in Wales', http://uknea.unep-wcmc.org/Resources/tabid/82/Default.aspx

10. P. J. Johnes et al, 2007, 'Land use scenarios for England and Wales: evaluation of management options to support "good ecological status" in surface freshwaters', *Soil Use and Management*, vol. 23 (suppl. 1), pp. 176–94.

11. UK National Ecosystem Assessment, 'Condition of a) riverine species, and b) riverine habitats in special areas of conservation in Wales'.

12. UK National Ecosystem Assessment, chap. 20, fig. 20.11, 'Threats to biodiversity in Wales', http://uknea.unep-wcmc.org/Resources/tabid/82/Default.aspx

13. Nigel Miller, vice-president of the National Farmers' Union in Scotland, 2008, quoted in LISS Online, www.oatridge.ac.uk/documents/982

14. Emyr Jones, 26 October 2012, letter to the *County Times*, Powys.

15. Chap. 20, http://uknea.unep-wcmc.org/Resources/tabid/82/Default.aspx.

16. UK National Ecosystem Assessment, chap. 20, fig. 20.31.

17. Statistics for Wales, Welsh Assembly government, 2010, *Farming Facts and figures, Wales.*

18. UK National Ecosystem Assessment, chap. 20, fig. 20.39, 'Imports and exports of food commodities in Wales', http://uknea.unep-wcmc.org/Resources/tabid/82/Default.aspx

19. UK National Ecosystem Assessment, chap. 20, fig. 20.22, 'Flood events in the River Wye from 1923 to 2003', http://uknea.unep-wcmc.org/Resources/tabid/82/Default.aspx

20. UK National Ecosystem Assessment, chap. 13, fig. 13.14, 'a) Long-term rainfall; and b) water balance (evapotranspiration) from the forested (Severn) and moorland (Wye) catchments at Plynlimon', http://uknea.unep-wcmc.org/Resources/tabid/82/Default.aspx

21. UK National Ecosystem Assessment, chap. 20, http://uknea.unep-wcmc.org/Resources/tabid/82/Default.aspx

22. Ibid.

23. UK National Ecosystem Assessment, chap. 22, fig. 22.2, 'Economic values that would arise from a change of land use from farming to multi-purpose woodland in Wales (£ per year)', http://uknea.unep-wcmc.org/Resources/tabid/82/Default.aspx

24. Institute of Biological, Environmental and Rural Sciences, Aberystwyth University, 2011, 'Farm outputs – all sizes. Table B3: Hill sheep farms, 2009/2010', http://www.aber.ac.uk/en/media/0910Iy_11d.pdf
25. DEFRA press office, 26 November 2011, by email.
26. DEFRA press office, 31 August 2011, by email.
27. Office of National Statistics, 2010, *Family Spending 2010 Edition*. *Table A1: Components of Household Expenditure 2009*, http://www. ons.gov.uk/ons/publications/re-reference-tables.html?edition=tcm%3A77-225698
28. Ibid.
29. Statistics for Wales, *Agricultural Small Area Statistics for Wales, 2002 to 2010*.
30. Official Journal of the European Union, 31 January 2009, 'Council Regulation (EC) No. 73/2009 of 19 January 2009, establishing common rules for direct support schemes for farmers under the common agricultural policy and establishing certain support schemes for farmers, amending Regulations (EC) No. 1290/2005, (EC) No. 247/2006, (EC) No. 378/2007 and repealing Regulation (EC) No. 1782/2003. Annex III', http://eur-lex.europa.eu/LexUriServ/LexUriServ.do?uri=OJ:L:2009: 030:0016:0016:EN:PDF
31. Miles King, December 2010, *An Investigation into Policies Affecting Europe's Semi-Natural Grasslands*, The Grasslands Trust, www. grasslands-trust.org/uploads/page/doc/European%20grasslands%20 report%20phase%201%20final%281%29.pdf
32. BBC Northern Ireland, 19 October 2011, 'Northern Ireland faces more European farm subsidy fines', www.bbc.co.uk/news/uk-northern-ireland-15369709; Miles King, 2011, 'Dark days return: farm subsidies drive environmental destruction', http://milesking.wordpress.com/2011/03/09/ dark-days-return-farm-subsidies-drive-environmental-destruction/; King, *Europe's Semi-Natural Grasslands*.
33. European Commission, 2011, 'Common Agricultural Policy towards 2020: Assessment of Alternative Policy Options', http://ec.europa.eu/ agriculture/analysis/perspec/cap-2020/impact-assessment/full-text_en.pdf
34. Welsh Assembly government, 2010, *Glastir: A Guide to Frequently Asked Questions*.
35. Ibid.
36. Welsh Assembly government, 2010, *Glastir Targeted Element: An Explanation of the Selection Process*.
37. See http://maps.forestry.gov.uk/imf/imf.jsp?site=fcwales_ext&
38. Genesis 1, 26.

39. Charlemagne, 30 October 2008, 'Europe's baleful bail-outs', http://www.economist.com/node/12510261

40. http://maps.forestry.gov.uk/imf/imf.jsp?site=fcwales_ext

41. Scottish Executive, Environment and Rural Affairs Department, 2007, *ECOSSE: Estimating Carbon in Organic Soils Sequestration and Emissions*, http://www.scotland.gov.uk/Publications/2007/03/16170508/16

42. Scottish Executive, Environment and Rural Affairs Department, 2007, as above.

43. James Morison et al, October 2010, 'Understanding the GHG implications of forestry on peat soils in Scotland', Forest Research, for Forestry Commission, Scotland, http://www.forestry.gov.uk/pdf/FCS_forestry_peat_GHG_final_Oct13_2010.pdf/$FILE/FCS_forestry_peat_GHG_final_Oct13_2010.pdf

44. European Commission, 'Common Agricultural Policy towards 2020'.

45. Environment Agency, 2009, 'Investing for the future: flood and coastal risk management in England', http://knowledgehub.local.gov.uk/c/document_library/get_file?uuid=ef1cd8ec-861d-4dd4-8518-6a59fc91ee1c&groupId=5919398

46. See National Trust Wales, 2008, 'Nature's capital: investing in the nation's natural assets', www.assemblywales.org/cr-lu2_natures_capital_wales_final.pdf

47. BBC Wales, 10 June 2012, 'Wales flooding: victims hoping for return to homes', http://www.bbc.co.uk/news/uk-wales-18384666; BBC Wales, 10 June 2012, 'Flood-risk villagers return home to Pennal in Gwynedd', http://www.bbc.co.uk/news/uk-wales-18387520

48. Wales Rural Observatory, 2007, *Population Change in Rural Wales: Social and Cultural Impacts. Research Report no. 14*, www.walesruralobservatory.org.uk/reports/english/MigrationReport_Final.pdf

10. THE HUSHINGS

1. Stephen Moss, 2012, *Natural Childhood*, The National Trust, www.nationaltrust.org.uk/servlet/file/store5/item823323/version1/Natural%20Childhood%20Brochure.pdf

2. See, for example, George Monbiot, 28 June 2010, 'A modest proposal for tackling youth', www.monbiot.com/2010/06/28/a-modest-proposal-for-tackling-youth/

3. Jay Griffiths, 2013, *Kith: The Riddle of the Childscape*, Hamish Hamilton. (I read the proof copy.)

4. George Monbiot, 1994, *No Man's Land: An Investigative Journey through Kenya and Tanzania*, Macmillan, London.

5. Richard Louv, 2009, *Last Child in the Woods*, Atlantic Books, London.

6. Andrea Faber Taylor, Frances E. Kuo and William C. Sullivan, 2001, 'Coping with ADD: the surprising connection to green play settings', *Environment and Behavior*, vol. 33, no. 1, pp. 54–77, doi: 10.1177/00139160121972864.

7. Robert Pyle, 2002, 'Eden in a vacant lot: special places, species and kids in community of life', in P. H. Kahn and S. R. Kellert (eds.), *Children and Nature: Psychological, Sociocultural and Evolutionary Investigations*, MIT Press, Cambridge, MA. Cited by Aric Sigman, no date given, 'Agricultural literacy: giving concrete children food for thought', http://www.face-online.org.uk/resources/news/Agricultural%20Literacy.pdf

8. G. A. Lieberman and L. Hoody, 1998, 'Closing the achievement gap: using the environment as an integrating context for learning', Sacramento, CA, CA State Education and Environment Roundtable, 1998, www.seer.org/pages/research. Cited by Sigman, 'Agricultural literacy'.

9. Simon Jenkins, 1 September 2011, 'If Britain fails to protect its heritage we'll have nothing left but ghosts', *Guardian*, http://www.guardian.co.uk/commentisfree/2011/sep/01/britain-industrial-heritage-dylife-wales

10. William Cronon, 1995, 'The trouble with wilderness; or, getting back to the wrong nature', in William Cronon (ed.), *Uncommon Ground: Rethinking the Human Place in Nature*, W. W. Norton & Co., New York, pp. 69–90.

11. R. Rasker and A. Hackman, 1996, 'Economic development and the conservation of large carnivores', *Conservation Biology*, vol. 10, pp. 991–1002.

12. S. Charnley, R. J. McLain and E. M. Donoghue, 2008, 'Forest management policy, amenity migration and community well-being in the American West: reflections from the Northwest Forest Plan', *Human Ecology*, vol. 36, pp. 743–61, doi: 10.1007/s10745-008-9192-3.

13. Kevin Cahill, 2002, *Who Owns Britain*, Canongate.

14. Department for Environment Food and Rural Affairs, January 2011, UK response to the Commission communication and consultation: 'The CAP towards 2020: meeting the food, natural resources and territorial challenges of the future', http://archive.defra.gov.uk/foodfarm/policy/capreform/documents/110128-uk-cap-response.pdf

15. Elizabeth Taylor, 16 November 2012, 'Heeding the coyote's call: Jim Sterba on the fight with wildlife over space in the sprawl', *Chicago Tribune*, http://articles.chicagotribune.com/2012-11-16/features/ct-prj-1118-book-of-the-month-20121116_1_wild-animals-wildlife-wild-game-meat/2

16. The Institute for European Environmental Policy, cited by Rewilding Europe, 2012, *Making Europe a Wilder Place*, www.rewildingeurope.com/assets/uploads/Downloads/Rewilding-Europe-Brochure-2012.pdf

11. THE BEAST WITHIN
(OR HOW NOT TO REWILD)

1. http://www.state.gov/r/pa/ei/bgn/3407.htm
2. http://en.wikipedia.org/wiki/Wales#Economy
3. W. H. Auden, 1965, 'Et in Arcadia Ego'.
4. Institute for Research of Expelled Germans, 2011, 'The forced labour, imprisonment, expulsion, and emigration of the Germans of Yugoslavia', http://expelledgermans.org/danubegermans.htm
5. Institute for Research of Expelled Germans, 'The forced labour, imprisonment, expulsion, and emigration of the Germans of Yugoslavia'.
6. Oto Luthar (ed.), 2008, *The Land Between: A History of Slovenia*, Peter Lang.
7. K. Kris Hirst, 2008, 'Lost cities of the Amazon', *National Geographic*, http://archaeology.about.com/od/ancientcivilizations/ss/expedition_week_6.htm
8. Anna Roosevelt, 1989, 'Resource management in Amazonia before the Conquest: beyond ethnographic projection', *Advances in Economic Botany*, vol. 7, The New York Botanical Garden.
9. Michael J. Heckenberger et al, 2003, 'Amazonia 1492: pristine forest or cultural parkland?', *Science*, vol. 301, no. 5640, pp. 1710–14, doi: 10.1126/science.1086112; Michael J. Heckenberger et al, 2008, 'Pre-Columbian urbanism, anthropogenic landscapes, and the future of the Amazon', *Science*, vol. 321, no. 5893, pp. 1214–17, doi: 10.1126/science.1159769.
10. Heckenberger et al, 'Pre-Columbian urbanism'.
11. Ran Prieur, 2010, 'Beyond civilised & primitive', *Dark Mountain*, vol. 1, pp. 119–35.
12. Richard Nevle and Dennis Bird, 17 December 2008, Presentation to the American Geophysical Union, http://news.stanford.edu/pr/2008/pr-manvleaf-010709.html
13. Felisa A. Smith, 2010, 'Methane emissions from extinct megafauna', *Nature Geoscience*, 3, pp. 374–5, doi: 10.1038/ngeo877.
14. Simon Schama, 1996, *Landscape and Memory*, Fontana Press, London.
15. Ibid.
16. E. P. Thompson, 1977, *Whigs and Hunters: The Origin of the Black Act*, Penguin, London.
17. Richard Leakey, quoted by George Monbiot, 1994, *No Man's Land: An Investigative Journey through Kenya and Tanzania*, Macmillan, London.

18. BBC Four, 16 June 2011, *Unnatural Histories*, http://www.bbc.co.uk/programmes/bo11wzrc

19. Forty-Second Congress of the United States of America, 1871, Act Establishing Yellowstone National Park (1872), http://www.ourdocuments.gov/doc.php?flash=true&doc=45&page=transcript

20. Susan S. Hughes, 2000, 'The Sheepeater Myth of Northwestern Wyoming', *Plains Anthropologist*, vol. 45, no. 171, pp. 63–83.

21. Boria Sax, 1997, '"What is a Jewish Dog?" Konrad Lorenz and the cult of wildness', *Society and Animals*, vol. 5, no. 1; Martin Brüne, 2007, 'On human self-domestication, psychiatry, and eugenics', *Philosophy, Ethics, and Humanities in Medicine*, vol. 2, no. 21, doi: 10.1186/1747-5341-2-21.

22. Sax, '"What is a Jewish Dog?"'.

23. Ibid.

24. Ibid.

25. Terry Eagleton, 2005, *The English Novel*, Blackwell, Malden, MA and Oxford, UK.

26. Francis Wheen, 18 September 1996, 'Sir Jimmy and the apeman: calling a Spode a Spode', *Guardian*.

27. Kim Sengupta, 30 June 2000, 'Death of a maverick: millionaire zoo-keeper from another era who cut a swathe through British business loses three-year fight against cancer', *Independent*.

28. Alexander Chancellor, 25 November 2000, 'John Aspinall's unspeakable behaviour was of a kind that would have landed almost anyone else in prison, and yet, to some, he died a hero', *Guardian*.

29. No author given, 30 June 2000, 'Obituary: John Aspinall', *The Times*.

30. Ros Coward, 13 February 2000, 'Profile: John Aspinall', *Observer*.

31. Martin Bright, 9 January 2005, 'Desperate Lucan dreamt of fascist coup', *Observer*.

32. Caroline Cass, 1994, *Joy Adamson: Behind the Mask*, FA Thorpe, Anstey.

33. Ibid.

34. Jamie Lorimer and Clemens Driessen, 2011, 'Bovine biopolitics and the promise of monsters in the rewilding of Heck cattle', *Geoforum*, in press, doi: 10.1016/j.geoforum.2011.09.002.

35. T. van Vuure, 2002, 'History, morphology and ecology of the aurochs (*Bos primigenius*)', *Lutra*, vol. 45, no. 1, pp. 1–16.

36. F. W. M. Vera, 2009, 'Large-scale nature development – the Oostvaardersplassen', *British Wildlife*, vol. 20, no. 5 (special supplement), pp. 28–36.

37. Lorimer and Driessen, 'Bovine biopolitics and the promise of monsters in the rewilding of Heck cattle'.

12. THE CONSERVATION PRISON

1. Noticeboard at the entrance of the reserve.
2. Montgomeryshire Wildlife Trust, 2009, *Glaslyn Management Plan 2009–2014*.
3. Montgomeryshire Wildlife Trust, 2010, *The Pumlumon Project. Two Year Progress Report 2008–2010*, http://www.montwt.co.uk/images/user/Pumlumon%20progress%20report%202010.pdf
4. Montgomeryshire Wildlife Trust, 2009. *Glaslyn Management Plan 2009–2014*.
5. Joint Nature Conservation Committee, 2004, *Common Standards Monitoring: Introduction to the Guidance Manual*, http://jnce.defra.gov.uk/pdf/CSM_introduction.pdf
6. Joint Nature Conservation Committee, 2009, *Common Standards Monitoring Guidance for Upland Habitats*, http://jnce.defra.gov.uk/pdf/CSM_Upland_jul_09.pdf
7. The European Habitats Directive, 21 May 1992, 'Council Directive 92/43/EEC of 21 May 1992 on the conservation of natural habitats and of wild fauna and flora', http://eur-lex.europa.eu/LexUriServ/LexUriServ.do?uri-CELEX:31992L0043:EN:HTML
8. Joint Nature Conservation Committee, 2012, *UK Interest Features*, http://jnce.defra.gov.uk/Publications/JNCC312/UK_habitat_list.asp
9. Joint Nature Conservation Committee, 2007, *Species and Habitats Review*, http://jnce.defra.gov.uk/PDF/UKBAP_Species+HabitatsReview-2007.pdf
10. British Trust for Ornithology and Joint Nature Conservation Committee, 2012, 'Red Grouse', http://www.bto.org/birdtrends2010/wcrredgr.shtml; British Trust for Ornithology and Joint Nature Conservation Committee, 2012, 'Skylark', http://www.bto.org/birdtrends2004/wcrskyla.htm; British Trust for Ornithology and Joint Nature Conservation Committee, 2012, 'Wheatear', http://www.bto.org/birdtrends2010/wcrwheat.shtml; British Trust for Ornithology, 2012, 'Ring Ouzel', http://blx1.bto.org/birdfacts/results/bob11860.htm
11. Patrick Barkham, 14 September 2011, 'Record numbers of golden eagles poisoned in Scotland in 2010', http://www.guardian.co.uk/environment/2011/sep/14/golden-eagles-poisoned-scotland-rspb
12. Severin Carrell, 27 May 2011, 'Gamekeeper with huge cache of bird poison fined £3,300', http://www.guardian.co.uk/uk/2011/may/27/gamekeeper-banned-pesticide-fined

13. Montgomeryshire Wildlife Trust, no date given, Newsletter, http://www.montwt.co.uk/newsletter/grouse%20count%20detail%20article%20final.htm

14. Estelle Bailey, Montgomeryshire Wildlife Trust, 17 June 2011, by email.

15. Powys County Council, 2011, *Upland and Lowland Heath Action Plan,* http://www.powys.gov.uk/uploads/media/upland_lowland_heath_bi.pdf

16. Frans Vera, 2000, *Grazing Ecology and Forest History,* CABI Publishing, Wallingford.

17. J. H. B. Birks, 2005, 'Mind the gap: how open were European primeval forests?', *Trends in Ecology & Evolution,* vol. 20, pp. 154–6; R. Fyfe, 2007, 'The importance of local-scale openness within regions dominated by closed woodland', *Journal of Quaternary Science,* vol. 22, no. 6, pp. 571–8, doi: 10.1002/jqs.1078.

18. Oliver Rackham, 2003, *Ancient Woodland: Its History, Vegetation and Uses in England,* Castlepoint Press, Dalbeattie. Cited in Kathy H. Hodder et al, 2009, 'Can the pre-Neolithic provide suitable models for re-wilding the landscape in Britain?', *British Wildlife,* vol. 20, no. 5 (special supplement), pp. 4–15.

19. N. J. Whitehouse and D. Smith, 2010, 'How fragmented was the British Holocene wildwood? Perspectives on the "Vera" grazing debate from the fossil beetle record', *Quaternary Science Reviews,* vol, 29, nos. 3–4, pp. 539–53, doi.org/10/1016/j/quascirev. 2009.10.010.

20. J. C. Svenning, 2002, 'A review of natural vegetation openness in north-western Europe', *Biological Conservation,* vol. 104, pp. 133–48.

21. R. H. W. Bradshaw, G. E. Hannon and A. M. Lister, 2003, 'A long-term perspective on ungulate-vegetation interactions', *Forest Ecology and Management,* vol. 181, pp. 267–80.

22. F. J. G. Mitchell, 2005, 'How open were European primeval forests? Hypothesis testing using palaeoecological data', *Journal of Ecology,* vol. 93, pp. 168–77; Hodder et al, 'Can the pre-Neolithic provide suitable models for re-wilding the landscape in Britain?'

23. Clive Hambler and Susan M. Canney, 2013 (2nd edn), *Conservation,* Cambridge University Press, Cambridge (read in galley proof).

24. P. Shaw and D.B.A. Thompson, 2006. The nature of the Cairngorms: diversity in a changing environment. TSO: Edinburgh. 444 pp. ISBN: 9780114973261 http://www.tsoshop.co.uk/bookstore.asp?FO=116001 3&ProductID=9780114973261&Action=Book.

25. Montgomeryshire Wildlife Trust, 2010, Heather Moorland and Bog Habitat Action Plan, http://www.montwt.co.uk/Heathermoorlandand bogactionplan.html

26. Montgomeryshire Wildlife Trust, *Glaslyn Management Plan 2009–2014*.

27. See Heather Crump and Mick Green, 2012, 'Changes in breeding bird abundances in the Plynlimon SSSI 1984–2011', *Birds in Wales*, vol. 9, no. 1.

28. Montgomeryshire Wildlife Trust, *Glaslyn Management Plan 2009–2014*.

29. Dr Barbara Jones, Countryside Council for Wales, February 2007, *A Framework to Set Conservation Objectives and Achieve Favourable Condition in Welsh Upland SSSIs*, http://www.ccgc.gov.uk/PDF/UPland%20Framework%201.pdf

30. Clive Hambler and Martin Speight, 1995, 'Biodiversity conservation in Britain: science replacing tradition', *British Wildlife*, vol. 6, no. 3, pp. 137–48.

31. N. Noe-Nygaard, T. D. Price and S. U. Hede, 2005, 'Diet of aurochs and early cattle in southern Scandinavia: evidence from 15N and 13C stable isotopes', *Journal of Archaeological Science*, vol. 32, pp. 855–71, doi: 10.1016/j.jas.2005.01.004.

32. The Mammal Society, 2011, http://www.mammal.org.uk/index.php?option=com_content&view=article&id=250&Itemid=283.

33. Derek Yalden, 1999, *The History of British Mammals*, T and AD Poyser, London.

34. R. Coard and A. T. Chamberlain, 1999, 'The nature and timing of faunal change in the British Isles across the Pleistocene/Holocene transition', *The Holocene*, vol. 9, no. 3, pp. 372–6, doi: 10.1191/095968399672435429.

35. The Mammal Society, 2011.

36. Robert S. Sommer et al. 2011, 'Holocene survival of the wild horse in Europe: a matter of open landscape?', *Journal of Quaternary Science*, vol. 26, no. 8, pp. 805–12, doi: 10.1002/jps.1509.

37. The Mammal Society, 2011.

38. Hambler and Canney, *Conservation*.

39. John Lawton, 2010, *Making Space for Nature: A Review of England's Wildlife Sites and Ecological Network*, DEFRA, http://archive.defra.gov.uk/environment/biodiversity/documents/201009space-for-nature.pdf

40. Clive Hambler, Peter A. Henderson and Martin R. Speight, 2011, 'Extinction rates, extinction-prone habitats, and indicator groups in Britain and at larger scales', *Biological Conservation*, vol. 144, pp. 713–21, doi: 10.1016/j.biocon.2010.09.004.

41. Jones, *A Framework to Set Conservation Objectives*.

42. Gareth Browning and Rachel Oakley, 2009, 'Wild Ennerdale', *British Wildlife*, vol. 20, no. 5 (special supplement), pp. 56–8.

43. The Wildlife Trusts, 2009, 'A living landscape: a call to restore the UK's battered ecosystems, for wildlife and people', updated, http://www.wildlifetrusts.org/sites/wt-main.live.drupal.precedenthost.co.uk/files/A%20 Living%20Landscape%20report%202009%20update.pdf

44. See, for example, PAN Parks Foundation, 2009, *As Nature Intended: Best Practice Examples of Wilderness Management in the Natura 2000 Network*, http://wwf.panda.org/about_our_earth/?uNewsID=192724

13. REWILDING THE SEA

1. Sam Davis, 2008, *Spider Crabs – the Wildebeest of our Waters*, http:// helfordmarineconservation.co.uk/publications/newsletters/spider-crabs-the-wildebeest-of-our-waters/

2. Callum Roberts, 2007, *The Unnatural History of the Sea*, Gaia, London.

3. Oliver Goldsmith, 1776, *An History of the Earth and Animated Nature*, vol. VI, James Williams, Dublin. Cited by Roberts, *Unnatural History of the Sea*.

4. Mike Thrussell, 2010, 'History of the British tuna fishery', http://www.worldseafishing.com/features/britishtuna.html

5. Yorkshire Film Archive, no date given, 'Tunny in Action', http://www.yfaonline.com/film/tunny-action

6. Andrew Burnaby, quoted by Roberts, *Unnatural History of the Sea*.

7. Potomac Conservancy, 2012, 'Find out about Potomac water quality', http://www.potomac.org/site/water-quality/

8. Christopher Mitchelmore, 2010, 'Newfoundland & Labrador cod fishery', http://liveruralnl.com/2010/07/17/newfoundland-labrador-cod-fishery/

9. Roberts, *Unnatural History of the Sea*.

10. J. Roman and S. R. Palumbi, 2003, 'Whales before whaling in the North Atlantic', *Science*, vol. 301, no. 5632, pp. 508–10.

11. Roberts, *Unnatural History of the Sea*.

12. Fred Pearce, 9 June 2001, 'Who's the real killer?', *New Scientist*, http:// www.newscientist.com/article/mg17022942.600-whos-the-real-killer.html; Sidney Holt, 2003, 'The tortuous history of "scientific" Japanese whaling', *BioScience*. Cited by Joe Roman and James J. McCarthy, 2010, 'The whale pump: marine mammals enhance primary productivity in a coastal basin', *PLoS ONE*, vol. 5, no. 10, pp. 1–8, doi: 10.1371/journal.pone.0013255.

13. Stephen Nicol, 12 July 2011, 'Vital giants: why living seas need whales', *New Scientist*, http://www.newscientist.com/article/mg21128201.700-vital-giants-why-living-seas-need-whales.html

14. Stephen Nicol et al, 2010, 'Southern Ocean iron fertilization by baleen whales and Antarctic krill', *Fish and Fisheries*, vol. 11, pp. 203–9.
15. Kakani Katija and John O. Dabiri, 2009, 'A viscosity-enhanced mechanism for biogenic ocean mixing', *Nature*, vol. 460, pp. 624–7, doi: 10.1038/nature08207.
16. Nicol et al, 'Southern Ocean iron fertilization by baleen whales and Antarctic krill'.
17. Roman and McCarthy, 'The whale pump: marine mammals enhance primary productivity in a coastal basin'.
18. Daniel G. Boyce, Marlon R. Lewis and Boris Worm, 2010, 'Global phytoplankton decline over the past century', *Nature*, vol. 466, pp. 591–6, doi: 10.1038/nature09268.
19. Nichol, 'Vital giants: why living seas need whales'.
20. Trish J. Lavery et al, 2010, 'Iron defecation by sperm whales stimulates carbon export in the Southern Ocean', *Proceedings of the Royal Society: B*, vol. 277, pp. 3527–31, doi: 10.1098/rspb.2010.0863.
21. A. J. Pershing et al, 2010, 'The impact of whaling on the ocean carbon cycle: why bigger was better', *PLoS One*, vol. 5, e12444. Cited by James A. Estes et al, 2011, 'Trophic downgrading of planet Earth', *Science*, vol. 333, pp. 301–6, doi: 10.1126/science.1205106.
22. Ransom A. Myers et al, 2007, 'Cascading effects of the loss of apex predatory sharks from a coastal ocean', *Science*, vol. 315, pp. 1846–50, doi: 10.1126/science.1138657.
23. Ibid.
24. Julia K. Baum and Boris Worm, 2009, 'Cascading top-down effects of changing oceanic predator abundances', *Journal of Animal Ecology*, vol. 78, pp. 699–714, doi: 10.1111/j.1365-2656.2009.01531.x.
25. Ibid.
26. Friedrich W. Köster and Christian Möllmann, 2000, 'Trophodynamic control by clupeid predators on recruitment success in Baltic cod?', *ICES Journal of Marine Science*, vol. 57, pp. 310–23, doi: 10.1006/jmsc.1999.0528.
27. The Royal Commission on Environmental Pollution, 2004, *Turning the Tide: Addressing the Impact of Fisheries on the Marine Environment, 25th Report*.
28. Jeremy B. C. Jackson et al, 2001, 'Historical overfishing and the recent collapse of coastal ecosystems', *Science*, vol. 293, pp. 629–38.
29. James A. Estes and David O. Duggins, 1955, 'Sea otters and kelp forests in Alaska: generality and variation in a community ecological paradigm', *Ecological Monographs*, vol. 65, no. 1, pp. 75–100.

30. Shauna E. Reisewitz, James A. Estes and Charles A. Simenstad, 2006, 'Indirect food web interactions: sea otters and kelp forest fishes in the Aleutian archipelago', *Oecologia*, vol. 146, pp. 623–31, doi: 10.1007/s00442-005-0230-1; Jackson et al, 'Historical overfishing and the recent collapse of coastal ecosystems'.

31. Ole Theodor Oslen, 1883, *The Piscatorial Atlas of the North Sea, English Channel, and St. George's Channels*, Grimsby.

32. Jackson et al, 'Historical overfishing and the recent collapse of coastal ecosystems'.

33. Ibid.

34. Ibid.

35. Georgi M. Daskalov, 2002, 'Overfishing drives a trophic cascade in the Black Sea', *Marine Ecology Progress Series*, vol. 225, pp. 53–63.

36. C. P. Lynam et al, 2011, 'Have jellyfish in the Irish Sea benefited from climate change and overfishing?', *Global Change Biology*, vol. 17, no. 2, pp. 767–82, doi: 10.1111/j.1365-2486.2010.0235.

37. J. Molloy, 1975, *The Summer Herring Fishery in the Irish Sea in 1974*, Department of Agriculture and Fisheries, Ireland, http://oar.marine.ie/bistream/10793/493/1/Irish%20Fisheries%20Leaflet%20No%2070.pdf

38. Anthony J. Richardson et al, 2009, 'The jellyfish joyride: causes, consequences and management responses to a more gelatinous future', *Trends in Ecology & Evolution*, vol. 24, no. 6, pp. 312–22.

39. Ibid.

40. Ruth H. Thurstan, Simon Brockington and Callum M. Roberts, 2010, 'The effects of 118 years of industrial fishing on UK bottom trawl fisheries', *Nature Communications*, vol. 1, no. 15, pp. 1–6, doi: 10.1038/ncomms1013.

41. Some of these are listed in ibid.

42. *UK National Ecosystem Assessment: Synthesis of the Key Findings*, 2011, http://uknea.unep-wcmc.org/Resources/tabid/82/Default.aspx

43. *New Scientist*, 17 May 2003, 'Old men of the sea have all but gone', http://www.newscientist.com/article/mg17823950.200-old-men-of-the-sea-have-all-but-gone.html; see also Ransom Myers and Boris Worm, 2003, 'Rapid worldwide depletion of predatory fish communities', *Nature*, vol. 423, pp. 280–83.

44. Quoted in Roberts, *Unnatural History of the Sea*.

45. Royal Commission on Environmental Pollution, *Turning the Tide*.

46. Dan Jones, 19 November 2009, 'Scuba diving to the depths of human history', *New Scientist*, http://www.newscientist.com/article.mg20427351.000-scuba-diving-to-the-depths-of-human-history.html; Hampshire and Wight

Trust for Maritime Archaeology, 2011, http://www.hwtma.org.uk/bouldnor-cliff

47. John Vidal, 27 February 2012, 'Overfishing by European trawlers could continue if EU exemption agreed', *Guardian*, http://www.guardian.co.uk/environmental/2012/feb/27/overfishing-european-trawlers-eu-exemption

48. Ibid.

49. Severin Carrell, 24 February 2012, 'Fishing skippers and factory fined nearly £1m for illegal catches', *Guardian*, http://www.guardian.co.uk/environment/2012/feb/24/fishing-skippers-fined-illegal-catches

50. Justin McCurry, 26 March 2010, 'How Japanese sushi offensive sank move to protect sharks and bluefin tuna', *Guardian*, http://www.guardian.co.uk/environment/2010/mar/26/endangered-bluefin-tuna-sharks-oceans

51. Justin McCurry, 5 January 2012, 'Bluefin tuna fish sells for record £473,000 at Tokyo auction', *Guardian*, http://www/guardian.co.uk/world/2012/jan/05/japanese-half-million-pound-tuna

52. Royal Commission on Environmental Pollution, *Turning the Tide*.

53. Oceana, 2012, 'More on bottom trawling gear', http://oceana.org/en/our-work/promote-responsible-fishing/bottom-trawling/learn-act/more-on-bottom-trawling-gear

54. WWF, 2012, 'Fishing problem: destructive fishing practices', http://wwf.panda.org/about_our_earth/blue_planet/problems/problems_fishing/destructive_fishing/destructive_fishing/

55. Hanneke Van Lavieren, 2012, 'Can no-take fishery reserves help protect our oceans?', http://ourworld.unu.edu/en/can-no-take-fisheries-help-protect-our-oceans/

56. IUCN, 2003, 'World Parks Congress Recommendations', http://cms-data.iucn.org/downloads/recommendationen.pdf

57. Nicola Jones, 16 May 2011, 'Marine protection goes large', http://www.nature.com/news/2011/110516/full/news.2011.292.html

58. Royal Commission on Environmental Pollution, *Turning the Tide*.

59. Thomas Bell, 2012, '127 marine conservation zones', http://www.marinereservescoalition.org/2012/12/03/127-marine-conservation-zones/

60. Sarah E. Lester, 2009, 'Biological effects within no-take marine reserves: a global synthesis', *Marine Ecology Progress Series*, vol. 384, pp. 33–46, doi: 10.3354/meps08029.

61. Royal Commission on Environmental Pollution, *Turning the Tide*.

62. English Nature, 22 July 2005, 'Lundy lobsters bounce back in UK's first no-take zone', press release.

63. M. G. Hoskin et al, 2011, 'Variable population responses by large deca-pod crustaceans to the establishment of a temperate marine no-take zone', *Canadian Journal of Fishers and Aquatic Sciences*, vol. 68, pp. 185–200, doi: 10.1139/F10-143.

64. Richard Black, 16 July 2008, 'Fishing ban brings seas to life', http://news.bbc.co.uk/1/hi/7508216.stm

65. Roberts, *Unnatural History of the Sea.*

66. Ibid.

67. Van Lavieren, 'Can no-take fishery reserves help protect our oceans?'

68. Royal Commission on Environmental Pollution, *Turning the Tide.*

69. Ibid.

70. Rupert Crilly and Aniol Esteban, 2012, 'Jobs lost at sea: overfishing and the jobs that never were', New Economics Foundation, http://new economics.org/sites/neweconomics.org/files/Jobs_Lost_at_Sea.pdf

71. R. Watson and D. Pauly, 2001, 'Systematic distortions in world fisheries catch trends', *Nature*, vol. 414, pp. 534–6. Cited Roberts, in *Unnatural History of the Sea.*

72. Mark Fisher, 2006, 'No take zones – a maritime rewilding', http://www.self-willed-land.org.uk/articles/no_take.htm

73. Richard Benyon, 15 November 2011, Written Ministerial Statement on Marine Conservation Zones, http://www.defra.gov.uk/news/2011/11/15/wms-marine-conservation-zones/

74. Jean-Luc Solandt, Marine Conservation Society, 12 March 2012, by email.

75. Joint Nature Conservation Committee, 2010, *Establishing Fisheries Management Measures to Protect Marine Conservation Zones*, http://jnce.defra.gov.uk/PDF/MCZ_FisheriesManagementFactsheet.pdf

76. Welsh Assembly Government, 2011, 'Marine Conservation Zone Project, Wales'. *Newsletter*, 3. http://www.werh.org/documents/110927 marinemcznewsletter3en.pdf

77. Marine Conservation Society, 2010, 'Welsh Assembly Government's sea protection plans a "disgraceful let down", says marine charity', http://www.mcsuk.org/press/view/327

78. Joint Nature Conservation Committee, 2012, 'Special Areas of Conser-vation', http://jncc.defra.gov.uk/page-23

79. Cardigan Bay Special Area of Conservation (SAC), various dates, http://www.cardiganbaysac.org.uk/

80. Cardigan Bay Special Area of Conservation (SAC) Management Scheme, Section 6.17, http://www.cardiganbaysac.org.uk/pdf%20files/Cardigan_Bay_SAC_Management_Scheme_2008.pdf

81. Letter from John Taylor, director of policy, CCW, to Graham Rees, Department for Rural Affairs, Welsh Assembly Government, 22 January 2010, Scallop Dredging.

82. CEFAS, 2011, 'Fisheries Management', http://www.cefas.defra.gov.uk/our-services/fisheries-management.aspx

83. CEFAS, 2011, 'Fisheries Science Partnership', http://www.cefas.defra.gov.uk/our-services/fisheries-management/fisheries-science-partnership.aspx

84. Sally Williams, 25 January 2011, 'Whale-watchers thrilled by the mighty fin', *Western Mail*, http://www.Walesonline.co.uk/news/local-news/cardigan/2011/01/25/whale-watchers-thrilled-by-the-mighty-fin-91466-28047175/

85. Wildlife Extra, June 2011, '21 fin whales spotted in Irish Sea', http://www.wildlifeextra.com/go/news/FIN-WHALE-uk.html

86. No author given, 26 January 2010, 'Rare sighting of humpback whale breaching in Irish sea caught on camera', *Daily Mail*, http://www.dailymail.co.uk/news/article-1246137/Rare-sighting-humpback-whale-breaching-Irish-sea-caught-camera.html

87. Thrussell, 'History of the British tuna fishery'.

14. THE GIFTS OF THE SEA

1. British Trust for Ornithology, 2007, 'Bird Atlas species index – Corncrake', http://blxl.bto.org/atlases/CE-atlas.html

2. Bird Care, 2012, 'Corncrake', www.birdcare.com/bin/showsonb?corncrake

3. *Macbeth*, Act 2, Scene 4.

Index